Insecticide resistance: from mechanisms to management

Insecticide resistance: from mechanisms to management

Edited by

I. Denholm,

J.A. Pickett

and

A.L. Devonshire

Department of Biological and Ecological Chemistry
IACR-Rothamsted
Harpenden
UK

CABI *Publishing*
in association with
The Royal Society

CABI Publishing is a division of CAB International

CABI Publishing
CAB International
Wallingford
Oxon OX10 8DE
UK

CABI Publishing
10 E 40th Street
Suite 3203
New York, NY 10016
USA

Tel: +44 (0)1491 832111
Fax: +44 (0)1491 833508
Email: cabi@cabi.org

Tel: +1 (212) 481 7018
Fax: +1 212 686 7993
Email: cabi-nao@cabi.org

A catalogue record for this book is available from the British Library, London, UK.

Library of Congress Cataloging-in-Publication Data
Insecticide Resistance : from mechanisms to management / edited by
 I. Denholm, J.A. Pickett, and A.L. Devonshire.
    p. cm.
  Based on a Discussion Meeting held at the Royal Society, London,
in April 1998.
  Includes bibliographical references and index.
  ISBN 0-85199-367-2 (alk. paper)
  1. Insecticide resistance. I. Denholm, Ian. II. Pickett, J. A.
(John A.) III. Devonshire, Alan L.
  SB951.5.I63   1999
  632'.95042--dc21   99-17258
                                                  CIP

ISBN 0 85199 367 2

First printed 1999
Transferred to print on demand 2001

Published in association with:
The Royal Society
6 Carlton House Terrace
London
SW1Y 5AG

Printed and bound in the UK by Antony Rowe Ltd, Eastbourne

Contents

Preface

Resistance of insects and other arthropod pests to chemically-based control strategies, whether involving direct application of pesticides to crops or the augmentation of crop plant defences by genetic manipulation, is an increasing threat to coping with the expected rapid increase in world population. Furthermore, the science that has already come from this subject has tremendous promise for showing how other anthropogenic influences may affect the world's fauna and even ourselves. The attention currently given to both basic and applied aspects of this subject, particularly in terms of our future food supply and in preserving species diversity, makes recent developments in research on insect resistance to chemical control exciting in a broad context.

This book contains papers presented at a Royal Society Discussion Meeting held in London in April, 1998. Its timing is crucial because of increasing concerns over resistance risks posed by over-reliance on insecticides and the rapidly expanding use of transgenic technology, particularly the expression in plants of insecticidal toxins derived from the microbe *Bacillus thuringiensis* (Bt). Although the largely commercially driven introduction of Bt crops has tended to play down the likely development of resistance to these plants, papers that follow highlight the risks involved and the world-wide need for resistance management strategies along the lines of ones implemented for Bt cotton in the USA and Australia, and others developed to combat resistance to conventional pesticides.

The opportunity for these issues to be reported and commented on in the prestigious and unbiased setting of the Royal Society arose from a suggestion by Michael Elliott, CBE, D.Sc, FRS, who led the pioneering work that gave us the pyrethroid insecticides. We see the papers that follow as providing the most comprehensive account yet of what has been achieved and where contributions can be expected in the future. There is also a widely held sentiment within this volume that we must continue to take forward and exploit the scientific legacy of the late Roman Sawicki, FRS. Roman, like Michael Elliott, was recruited to Rothamsted by the British insect control pioneer Charles Potter, but while Michael and his colleagues were developing the synthetic pyrethroids, initially with biological input from Roman, he was himself helping to establish the new science of insecticide resistance (Denholm and Elliott, 1995). The subject is now international, with leading groups from all over the world providing papers for this publication. However, there is a major message advocated by Roman that we still need to accept fully: for every toxin introduced for pest control, resistance is likely to develop. A great deal has been lost by over-use and inappropriate deployment of pest control agents. Although significant enlightenment now exists in the public and industrial research communities with regard to pesticide resistance management, commercial considerations are still in danger of precipitating the loss of valuable chemical structural types and modes of action. It was therefore with some apprehension (e.g. Pickett, 1997) that many viewed the rapid commercialisation of the first genetically delivered, Bt-based toxicants.

It is apparent from the following papers that particularly dramatic progress has been achieved in identifying and sequencing genes conferring insecticide resistance, and in many cases the mutational events leading to resistance have now been characterised in detail. One practical benefit of this is that it is now possible in a number of species to diagnose individual mechanisms using biochemical and/or DNA markers, and to investigate how these interact in determining resistance phenotypes. Another exciting and expanding area of research is the application of new techniques to map the distribution of resistance genes in insect genomes. The availability of DNA markers especially has overcome previously severe constraints presented by species whose formal genetics are still poorly understood, such as heliothine bollworms. From a more applied standpoint, there has been substantial progress with developing strategies based on a sound knowledge of pest ecology, resistance and cross-resistance for the rational deployment of insecticides in order to reduce selection for specific resistance mechanisms. Success with implementing these strategies over large areas reflects a much improved level of cooperation between academics, politicians, crop consultants and industrialists in many countries. The development of such strategies to conserve the effectiveness of transgenic plants is still at an early stage, but it is clear from some of the papers presented that the potential for resistance to these agents exists and that considerable effort is being made to deploy them in ways (e.g. in conjunction with refuge crops) to minimise the risk of insecticide resistance developing.

Although the commercial exploitation of plant genetics for the control of pests is exciting and novel, some of the problems it poses are not new and still require effective, long-term solutions. Based on discussions at the Royal Society meeting, however, there is great optimism that resistance management strategies will be an integral part of future plans. This publication should provide invaluable information upon which to base such approaches.

References
Denholm, I.; Elliott, M.E. (1995). Roman Mieczyslaw Sawicki. *Bibliographic Memoirs of Fellows of the Royal Society* **1995**: 397-417.
Pickett, J.A. (1997). Protection racket. *Chemistry and Industry* **1 December 1997**: 956-957.

Note
This Preface is a revised and expanded version of that which appeared in the *Philosophical Transactions*.

J. A. Pickett, I. Denholm and A. L. Devonshire

The evolution of insecticide resistance in the peach–potato aphid, *Myzus persicae*

Alan L. Devonshire[1], **Linda M. Field**[1], **Stephen P. Foster**[1], **Graham D. Moores**[1], **Martin S. Williamson**[1] and **R. L. Blackman**[2]

[1]*Biological and Ecological Chemistry Department, IACR-Rothamsted, Harpenden, Hertfordshire AL5 2JQ, UK*
[2]*Department of Entomology, The Natural History Museum, London SW7 5BD, UK*

The peach–potato aphid *Myzus persicae* (Sulzer) can resist a wide range of insecticides, but until recently (1990) the only mechanism identified was the increased production of carboxylesterases (E4 or FE4), which cause enhanced degradation and sequestration of insecticidal esters. We have now identified two forms of target-site resistance involving changes in the acetylcholinesterase (*AChE*) and sodium channel (*kdr*) genes. Biochemical and DNA diagnostic methods can be used to identify all three mechanisms in individual aphids, and thereby establish their spatial distributions and temporal dynamics. Amplified genes underlie the increased production of esterases, but their expression is modulated by DNA methylation. Amplification of the *E4* gene is in strong linkage disequilibrium with the kdr mechanism. This may reflect strong insecticidal selection favouring aphids with multiple mechanisms, tight chromosomal linkage and/or the prominence of parthenogenesis in many *M. persicae* populations. The decreased fitness of resistant aphids under winter conditions may be a consequence of the altered sodium-channel gene affecting behaviour and/or the perception of external stimuli.

Keywords: aphids; carboxylesterases; chromosomal linkage; gene amplification; target-site resistance

1. GENETIC CONSIDERATIONS

Any consideration of the evolution of resistance in peach–potato aphids (*Myzus persicae*), must take into account their complex life cycle, which can vary according to the different environments in which they occur (Blackman 1974). They have herbaceous summer (secondary) hosts, which include many annual crops such as potatoes, sugar beet, chrysanthemums, tobacco and various brassicas, on which they reproduce by parthenogenesis. Parthenogenetic populations develop as a mixture of clones, with the most favoured ones as potential dominators of the population. They can then overwinter either asexually (through continued parthenogenesis on protected crops, winter weeds or field crops such as oil seed rape) or as sexually-produced eggs on a woody (primary) host. Because *M. persicae* has a strict requirement for peach (*Prunus persica*) as its primary host, sexual reproduction and meiotic recombination are confined to those areas where this tree is present. The variation in reproductive mode is not only due to the environment but is also dependent on genotype (Blackman 1974). Thus, some clonal lines respond to autumn conditions by producing males and sexual egg-laying females (holocycly), whereas others can only reproduce through continuous parthenogenesis (anholocycly), and yet others are able to produce males but not females (androcycly). Clearly, this will have a large impact on population structure and gene flow.

In addition to showing variation in their reproductive mode, *Myzus* can display preferences between summer hosts. This can be very marked, to the extent that the aphids become recognizable morphometrically, as with the tobacco-feeding form, which has been classified as a distinct species, *M. nicotianae* (Blackman 1987). Such segregation would again be expected to restrict gene flow. However, in areas where peach is grown close to tobacco crops, for example in Greece, there is the opportunity for genetic mixing because the two forms retain their ability to interbreed.

A third genetic feature relevant to the evolution of resistance in *M. persicae* is their karyotype. It has long been recognized that a heterozygous translocation between autosomes 1 and 3 is associated with one type of esterase-based resistance (Blackman *et al.* 1978). The translocation appears to reduce the aphid's ability to reproduce sexually, again having implications for the spread of resistance genes. Aphids with the A 1,3 translocation are widely distributed in warm temperate and tropical regions of the world, and in protected (glasshouse) crops in northern Europe, where they are mostly either anholocyclic or androcyclic.

2. INSECTICIDE DETOXIFICATION BY ESTERASES

(a) *Biochemistry*

For many years, the only resistance mechanism identified in *M. persicae* was the overproduction of insecticide-detoxifying esterases. This form of resistance was first implicated in the late 1960s by the demonstration that all resistant strains showed an increased ability to hydrolyse the model esterase substrate, 1-naphthyl acetate (Needham & Sawicki 1971). It was subsequently shown to arise from the increased production of one of two forms of esterase, E4 or FE4, that were distinguishable electrophoretically (reviewed by Devonshire 1989). These esterases can account for as much as one per cent of the aphid's total body protein, and give a broad spectrum of resistance to organophosphorus, carbamate and pyrethroid insecticides as a consequence of both ester hydrolysis and sequestration (Devonshire & Moores 1982). Although FE4 hydrolyses insecticides slightly faster than E4, the form of esterase overproduced has little effect on the level of resistance (Devonshire *et al*. 1983).

The biochemical evidence for the role of these esterases in resistance was supported by selection experiments: spraying *M. persicae* populations with pyrethroid insecticides selected strongly for aphids carrying esterase-based resistance to organophosphates and carbamates, and vice versa. This selection has been observed both in the open field (ffrench-Constant *et al*. 1988*a*) and in field cages colonized by known initial proportions of susceptible and resistant aphid clones (ffrench-Constant *et al*. 1987). However, for pyrethroids the contribution of esterases to resistance has recently been shown to be secondary to that conferred by target-site insensitivity, with the two mechanisms being co-selected as a consequence of a strong linkage disequilibrium between them (see below).

(b) *Amplification of esterase genes*

The observation that aphid clones with increased amounts of esterase fell into a series with a progressive doubling of enzyme activity, led to the hypothesis that the underlying cause of resistance was amplification of their structural genes, rather than transcriptional control (Devonshire & Sawicki 1979). This was confirmed in later work, but the phenomenon was shown to be more complex than a simple series of successive duplications (Field *et al*. 1988). The genes encoding the two esterase forms, E4 and FE4, each span approximately 5 kb and are very similar (Field *et al*. 1993). They differ by only nine amino-acid substitutions; and FE4 has an additional 12 amino acids at the C-terminus as a consequence of a mutation that changes the stop codon present in the *E4* gene. All clones studied so far have had identical *E4* or *FE4* genes, including the introns that have been sequenced. This suggests that a single amplification event occurred for each gene, and that they then became dispersed. Likewise, the tobacco-feeding form, *M. nicotianae*, has been shown to have exactly the same amplified genes, indicating that there is gene flow between this and *M. persicae* (Field *et al*. 1994).

In both *M. persicae* and *M. nicotianae*, amplification of *E4* genes always occurs in conjunction with the A 1,3 translocation, whereas aphids that have only the amplified *FE4* genes are apparently of normal karyotype. It is very rare to find both forms of the esterase genes amplified in a single individual, pointing to some reproductive isolation between aphids carrying the different genes. The two amplified genes were first seen together in the progeny of laboratory crosses between *E4* and *FE4* clones (Blackman *et al*. 1996); and the combination has recently been found in a small number of aphids collected in the field in Greece (Blackman *et al*. 1998). In the latter case, aphids with both genes amplified have also had the translocation.

The organization of the amplified genes has been studied by analysing large restriction fragments by means of pulsed-field gel electrophoresis (PFGE), which gave very different results for *E4* and *FE4* genes. Rare-cutting enzymes gave a single DNA fragment of 300–350 kb for amplified *E4* genes in R_3 aphids, whereas those producing comparable amounts of the FE4 enzyme gave a family of fragments (Field *et al*. 1996). The complexity of the FE4 pattern precluded detailed analysis, but partial digests of the amplified *E4* genes indicated that they were arranged in a tandem array of 12 copies, with the esterase gene being part of a 24 kb amplicon (Field & Devonshire 1997). However, recent work (Field *et al*., unpublished data) has shown that the number of gene copies for both *E4* and *FE4* genes is higher, and more in line with the level of esterase production.

Further analysis of the *E4* and *FE4* genes has provided evidence that in susceptible aphids they are in a head-to-tail arrangement with *E4* upstream of *FE4*, with about 19 kb of intervening sequence (Field & Devonshire 1998). Their close similarity suggests a recent duplication event, followed by limited divergence of the two forms. The subsequent amplifications of the *E4* or *FE4* genes must have involved separate events, each occurring once and then being selected by insecticide treatment and spread by migration (figure 1). The 5′ ends of the genes differ, suggesting that they are regulated in different ways. *FE4* appears to have a conventional promoter region, whereas *E4* has a CpG island (associated with a 1.7 kb insertion that is absent from the *FE4* gene) that may influence expression through changes in DNA methylation (Field & Devonshire 1998). This would be in accord with the observation that only *E4* genes appear to undergo reversion through changes in transcription (see below). The novel joint (NJ) in the DNA sequence generated during the *E4* amplification is just upstream of the transcription start site, and the *FE4* joint (FJ) generated by the initial duplication event leading to the divergence of *FE4* is only 3 bp away from the NJ. This suggests that this part of the chromosome might be prone to recombination, perhaps as a consequence of the initial duplication. The position of the NJ generated by the *FE4* amplification has not yet been identified.

(c) *Chromosomal location and inheritance of esterase genes*

Fluorescence *in situ* hybridization (FISH) has enabled the sites of esterase gene amplification to be identified. The situation is simplest for amplified *E4* genes, which, with one exception amongst those clones studied, have always been found at a single locus on the truncated autosome 3, close to the translocation break-point. In the one exceptional clone, the amplified *E4* genes were also found at two other loci on autosomes 2 and 5 (Blackman *et al*. 1995). In contrast, *FE4* genes have been found widely dispersed around the genome, with interclonal variation in the number of loci, some apparently being homozygous and others heterozygous in any one clone (Blackman *et al*. 1995, 1996). These results are in line with the PFGE data, which indicated a single block of *E4* amplicons, but several differently sized groups of *FE4* genes (Field *et al*. 1996). Crossing experiments, some coupled with FISH analysis, have shown that inheritance patterns are consistent with a single heterozygous locus of *E4* amplification

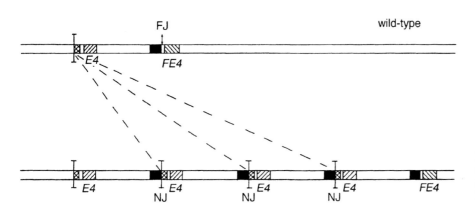

Figure 1. Model for the amplification and organization of *E4* and *FE4* genes (based on Field & Devonshire 1998).

in R_3 clones (Blackman *et al.* 1978, 1996). In contrast, two- or three-locus models were required for clones showing different degrees of *FE4* amplification (Blackman *et al.* 1996).

Despite their widespread dispersion around the genome, restriction analysis of the amplified genes indicates that all *FE4* copies are in the same immediate genetic background, i.e. the amplicon structures are maintained. The same conservation applies to the one aphid clone where the *E4* amplicons occur at three loci (Field & Devonshire 1998). This suggests that in each case an original amplification event occurred recently and the amplicon was then moved, intact, around the genome through rearrangements such as translocations and inversions, or perhaps mediated by transposable elements. The position and sequences around the site of initial duplication (FJ in figure 1) are identical in all of the aphid clones examined, reinforcing the likelihood of a single event that has, subsequently, become widely spread in populations around the world.

(d) *Transcriptional control of amplified esterase genes*

Some clonal cultures of aphids with amplified *E4*, but apparently not *FE4*, genes lose resistance spontaneously, concomitant with ceasing to over produce the esterase (Sawicki *et al.* 1980). This 'reversion' phenomenon occurs infrequently and stochastically in laboratory cultures in the absence of selection. However, recent evidence indicates that revertant aphids might be more common in the field than originally considered (L. M. Field and A. L. Devonshire, unpublished data). This loss of esterase-gene expression seems analogous to that seen in cell cultures resistant to cytotoxic drugs, which arises from loss of the amplified genes that confer resistance (Stark 1993). However, in *M. persicae*, revertant aphids retain their full complement of amplified esterase genes, suggesting that decreased transcription is responsible for the loss of enzyme production.

In aphids, the loss of expression was associated with changes in the methylation of the amplified esterase genes (Field *et al.* 1989); further work demonstrated a tight correlation between the two events (Hick *et al.* 1996). The amplified genes are expressed when 5-methylcytosine is present in and around the gene, and they are silenced when the methylation is lost in revertants. This positive correlation is contrary to the situation in vertebrates, where methylation is usually associated with gene silencing (Holliday *et al.* 1996). Revertant aphid clones spontaneously produce a small proportion of offspring with higher esterase levels, which are selected when exposed to insecticides (ffrench-Constant *et al.* 1988*b*). This plasticity confers a potential advantage to these aphid clones by enabling them to avoid producing the large amount of esterase protein when it is not needed.

(e) *Origin and spread of amplified esterase genes*

The molecular studies of amplified esterase genes, as in mosquitoes (Raymond *et al.*, this issue), points to a small number of isolated mutation events (Field & Devonshire 1997, 1998) which then spread rapidly through migration. The mode of aphid reproduction, coupled with their potential for rapid growth as asexual populations, mean that the parthenogenetic offspring of an individual with a marked advantage could become dominant very quickly. Furthermore, the migration of aphids, through both short- and long-distance flight, will lead to the dispersal of resistance genes. However, another way in which insects can become dispersed around the world is as a result of human activity. This has been invoked for mosquitoes, in which their movement through long-distance transport in boats and aircraft appears to have played a part in the rapid spread of resistance genes (Pasteur & Raymond 1996). Likewise, the international trade in plants and produce offers considerable potential for the widespread distribution of phytophagous insect pests (Frey 1993; Denholm *et al.*, this issue).

3. TARGET-SITE RESISTANCE

(a) *Insecticide-insensitive acetylcholinesterase*

Insecticide-insensitive acetylcholinesterase (AChE), the target for organophosphorus and carbamate insecticides, is an important resistance mechanism in many insect

pirimicarb

triazamate

aldicarb

methomyl

carbofuran

ethiofencarb

Figure 2. Chemical structures of N,N-dimethylcarbamates (pirimicarb and triazamate) to which the *Myzus* AChE is insensitive, and some N-monomethylcarbamates that are unaffected by this resistance mechanism.

species, but was not found in the *M. persicae – M. nicotianae* complex until 1990 (Moores *et al.* 1994*a*). It was first detected in populations from Greece, followed by those from Japan and South America, and has recently shown a northward expansion in its European distribution (Moores *et al.* 1994*b*). Aphids with insensitive AChE were first identified in the UK in 1995 from samples which were collected in aerial suction traps but did not cause any control problems in that year. However, in 1996, insensitive AChE occurred commonly in aphids from eastern England (Foster *et al.* 1998) where there were control failures with pirimicarb, the favoured insecticide for controlling the *M. persicae* that are very resistant (R_2 and R_3) owing to esterase overproduction.

The fact that insensitive AChE was only found recently (1990), and only in association with the esterase mechanism (both E4 and FE4), suggests that it evolved after the amplified esterases. The incidence of what appears to be the same *AChE* gene alongside both forms of esterase suggests that it originated in a holocyclic individual and then spread through the *M. persicae – M. nicotianae* complex. If it originated as a single mutation event, its introgression into aphids with the A 1,3 translocation (which are anholocyclic or androcyclic) must have been through a female mating with a male from an androcyclic E4-producing clone. The continuous parthenogenetic reproduction of the offspring of such a cross could then have enabled it to be stably maintained in a heterozygous condition, when its phenotypic expression (resistance to pirimicarb) is semi-dominant (Moores *et al.* 1994*b*). However, it has also been found in a homozygous

condition in combination with amplified *E4* as well as *FE4*, indicating further mating, presumably through *E4* males crossing with females of normal karyotype also with the insensitive *AChE* gene. In England, aphids with the insensitive *AChE* gene have all been red and with amplified *E4*, whereas in Greece, where *FE4* predominates, the insensitive form has been found in both the red and green colour morphs.

The insensitive AChE confers strong resistance specifically to pirimicarb and triazamate, the latter being a novel and otherwise very effective triazole aphicide (Dewar *et al.* 1994) which also inhibits AChE. Both are N,N-dimethylcarbamates; there is no insensitivity to a range of organophosphorus compounds nor to aryl and oxime N-monomethylcarbamates (figure 2), suggesting that the bulkier substituents on the carbamate nitrogen are critical for the expression of resistance in this particular species. The insensitivity appears to arise at the inhibition stage, rather than from differential reactivation. Reactivation rates (k_3, measured as described by Devonshire & Moores (1982)) were too slow to be determined for the sensitive AChE; whilst the insensitive form gave values of only $0.0029 \pm 0.0001\,\mathrm{min}^{-1}$ and $0.00047 \pm 0.00007\,\mathrm{min}^{-1}$ for mono- and dimethylcarbamylated enzyme, respectively.

Although the molecular basis of AChE insensitivity in other species involves one or more point mutations that change amino-acid residues close to the catalytic site of the enzyme (Mutero *et al.* 1994; Devonshire *et al.* 1998), the changes in *M. persicae* have not yet been established.

(b) *Modified sodium channels (knockdown resistance or kdr)*

We have recently shown that the esterase-based resistance to pyrethroids in *M. persicae* is of secondary importance compared with a kdr-type mechanism not previously identified in this species (Martinez-Torres *et al.* 1997, 1998), and that their co-selection appears to have been a consequence of their close association in some clones (Field *et al.* 1997). Aphids lacking the kdr mechanism, but with R_3 levels of esterase, only show approximately five-fold resistance to deltamethrin. The kdr mechanism alone (as found only in revertant aphids) confers 35-fold resistance and this is enhanced a further 15-fold by R_3 levels of esterase (figure 3). Analysis of aphid clones collected over many years, and reared in the laboratory, indicated that this mechanism had long been present in populations, but was only identified once a molecular diagnostic method became available. The cross-resistance it confers to DDT has since enabled the use of a discriminating dose bioassay for identifying its presence. The kdr mechanism involves a mutation causing the replacement of a leucine by phenylalanine in the domain IIS6 transmembrane region of the insect *para*-type sodium-channel gene. This mutation was first identified in house flies (Williamson *et al.* 1996) and cockroaches (Miyazaki *et al.* 1996), and the same change has since been found in several insect species including *M. persicae* (Martinez-Torres *et al.* 1997; ffrench-Constant *et al.*, this issue).

In UK populations, the kdr mechanism shows strong linkage disequilibrium with the amplification of *E4*, but not *FE4*, genes (table 1). This perhaps reflects the almost

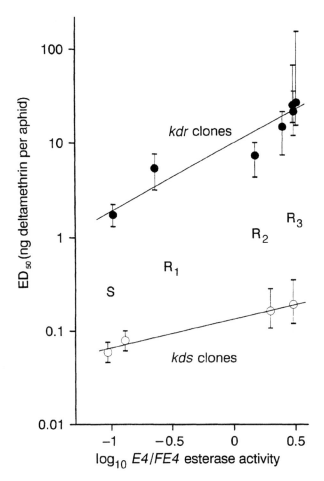

Figure 3. Relationship between toxicity of deltamethrin (ED_{50}) and mean esterase activity, measured by immunoassay, in *kdr* and *kds* (i.e. lacking the kdr mechanism) aphid clones. Limits of bars indicate 95% confidence intervals given by probit analysis. All of the *kdr* clones have highly amplified *E4* genes, but the clone with S esterase activity is a revertant, and those with R_1 and R_2 esterase levels appear to be partial revertants, based on analysis of DNA methylation patterns (Field *et al.* 1989).

exclusive chromosomal location of the amplified *E4* genes at a single locus on the truncated chromosome 3 close to the translocation break-point, whereas amplified *FE4* genes are more dispersed around the genome (Blackman *et al.* 1995, 1996, 1998). However, this linkage of the kdr mechanism with *E4* gene amplification does not hold fully outside the UK: in a broader survey of aphids from 15 countries, 5 of the 30 clones with amplified *E4* genes lacked the kdr mechanism (Field *et al.* 1997). It is not known whether the close association of the E4 and kdr mechanisms reflects a tight chromosomal linkage. It seems more likely that the *kdr* mutation first occurred in a single E4-overproducing, translocated aphid, and gave its progeny such a large selective advantage when exposed to a wide variety of insecticides that they survived preferentially and have become widely dispersed as a clone. Following this hypothesis, males from this (androcyclic) clone would then have passed the *kdr* mutation via the sexual phase to some FE4-overproducing genotypes.

Table 1. *Incidence of the kdr mechanism*[a] *in UK aphids with different esterase levels and amplification of* E4 *and* FE4 *genes*

| esterase level | number of clones with: | | | |
| | amplified *E4* | | amplified *FE4* | |
	kdr	kds[b]	kdr	kds
R_1	8	0	1	0
R_2	26	0	3	11
R_3	67	0	1	1

[a] Based on a diagnostic bioassay with DDT (Field *et al.* 1997).
[b] The corresponding susceptible phenotype is referred to as kds for simplicity.

4. RELATIVE FITNESS OF SUSCEPTIBLE AND RESISTANT APHIDS

One of the tenets of resistance management is that, in the absence of insecticides, resistant insects are less 'fit' than their susceptible counterparts. Early studies of *M. persicae* provided evidence for some resistant clones moving less readily between different host species in the laboratory (Eggers-Schumacher 1983). Evidence of a more marked fitness deficit under field conditions began to emerge from the results of regular monitoring of populations throughout and between growing seasons in England (Smith *et al.* 1990). The proportion of aphids with esterase-based resistance built up during the summer in populations sampled from crops and, more randomly, in the aerial suction traps of the Rothamsted Insect Survey, presumably reflected selection pressure from insecticide use on the various crop hosts of *M. persicae*. However, their proportions declined over several winters, so that the overall situation remained relatively stable from 1988 to 1995 (Muggleton *et al.* 1996), suggesting that the more resistant forms were less able to survive the winter climate.

The poorer overwintering performance of E4-over-producing resistant aphids, compared with that of susceptible aphids, was clearly demonstrated in four out of nine field experiments over the winters of 1992–3 and 1993–4 (Foster *et al.* 1996). Several distinct clones from each of the resistance categories, S, R_1, R_2, R_3 and revertants, were colonized as early-instar nymphs on to oilseed rape crops, and their survival determined a month later; one such trial is shown in figure 4. These predominantly showed a negative correlation between survival and resistance level, with revertants behaving similarly to their R_3 counterparts (indicating that overproduction of esterase *per se* is not responsible for the differences observed). The relationship correlated strongly with three meteorological variables: day temperature below 2 °C, mean rainfall and mean wind speed. One of the factors implicated in this maladaptive behaviour was the tendency for the more resistant aphids to remain on deteriorating leaves, whereas the more susceptible clones moved more readily. This was apparent in both field trials and laboratory experiments (Foster *et al.* 1997), and would be expected to occur to increase the risk of the resistant aphids being separated from the plant after leaf-fall. Another characteristic was that some resistant

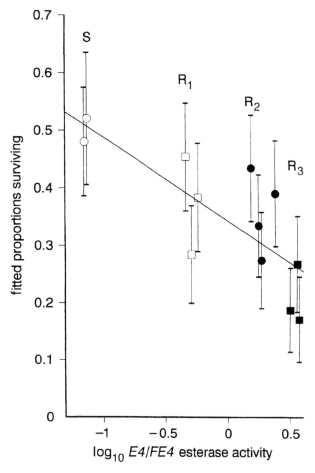

Figure 4. Relationship between esterase activity and the proportion of aphids surviving on oilseed rape crops in the field during a three week period in February 1994 (based on Foster *et al*. 1996). Symbols: open circles, S clones; open squares, R_1 clones; filled circles, R_2 clones; filled squares, R_3 clones. Limits of bars indicate ± 1 s.e.

clones of *M. persicae* were shown to be less responsive to their own alarm pheromone (Dawson *et al*. 1983); and more recent work (Foster *et al*., unpublished data) has shown this to be so in a wider range of clones.

At the time of these experiments, classification of resistant aphids was based on their esterase content. We now know that the kdr factor commonly occurs in combination with high E4 production, in UK clones. Because the mutation underlying kdr is in the sodium-channel gene, this would seem to be a more likely factor mediating the modified behaviour responsible for the reduced overwintering success, e.g. by reducing the sensitivity of the nervous system to stimuli. The *smellblind* mutant in *Drosophila melanogaster* provides a precedent for such an effect of a sodium-channel mutation (Lilly *et al*. 1994).

5. THE FUTURE

The way in which resistance will evolve in the future will depend on the pest control measures adopted. Our increasing knowledge of the underlying mechanisms, and the availability of sensitive and rapid diagnostic methods for their identification, opens the way to make rational choices of insecticides to minimize selection pressure in particular circumstances. Increasing pressure to minimize pesticide use is also likely to help reduce the selection for resistance. However, one of the immediate needs is for novel chemicals circumventing, or even exploiting (Hedley *et al*. 1998), existing mechanisms. The recent introduction of the chloronicotinyl class of aphicides, with its distinctive chemistry and mode of action, holds great promise for addressing this need (Elbert *et al*. 1990). However, there is already evidence for a low level of resistance to these compounds in some *M. persicae* populations (Devine *et al*. 1996), and if their promise is to be fulfilled, it will be essential to manage their use carefully from the outset. This issue is addressed by others in this volume.

We thank colleagues in the Resistance Group for their support and useful discussions throughout the work. IACR-Rothamsted receives grant-aided support from the Biotechnology and Biological Sciences Research Council of the United Kingdom.

REFERENCES

Blackman, R. L. 1974 Life cycle variation of *Myzus persicae* (Sulz.) (Hom., Aphididae) in different parts of the world, in relation to genotype and environment. *Bull. Entomol. Res.* **63**, 595–607.

Blackman, R. L. 1987 Morphological discrimination of a tobacco-feeding form from *Myzus persicae* (Sulzer) (Hemiptera: Aphididae), and a key to New World *Myzus* (*Nectarosiphon*) species. *Bull. Entomol. Res.* **77**, 713–730.

Blackman, R. L., Takada, H. & Kawakami, K. 1978 Chromosomal rearrangement involved in insecticide resistance of *Myzus persicae*. *Nature* **271**, 540–542.

Blackman, R. L., Spence, J. M., Field, L. M. & Devonshire, A. L. 1995 Chromosomal location of the amplified esterase genes conferring resistance to insecticides in the aphid *Myzus persicae*. *Heredity* **75**, 297–302.

Blackman, R. L., Spence, J. M., Field, L. M., Javed, N., Devine, G. J. & Devonshire, A. L. 1996 Inheritance of the amplified esterase genes responsible for insecticide resistance in *Myzus persicae* (Homoptera: Aphididae). *Heredity* **77**, 154–167.

Blackman, R. L., Spence, J. M., Field, L. M. & Devonshire, A. L. 1998 Variation in the chromosomal distribution of amplified esterase (FE4) genes in Greek field populations of *Myzus persicae* (Sulzer). *Heredity*. (In the press.)

Dawson, G. W., Griffiths, D. G., Pickett, J. A. & Woodcock, C. M. 1983 Decreased response to alarm pheromone by insecticide-resistant aphids. *Naturwissenschaften* **70**, 254–255.

Devine, G. J., Harling, Z. K., Scarr, A. W. & Devonshire, A. L. 1996 Lethal and sublethal effects of imidacloprid on nicotine-tolerant *Myzus nicotianae* and *Myzus persicae*. *Pestic. Sci.* **48**, 57–62.

Devonshire, A. L. 1989 Insecticide resistance in *Myzus persicae*: from field to gene and back again. *Pestic. Sci.* **26**, 375–382.

Devonshire, A. L. & Sawicki, R. M. 1979 Insecticide-resistant *Myzus persicae* as an example of evolution by gene duplication. *Nature* **280**, 140–141.

Devonshire, A. L. & Moores, G. D. 1982 A carboxylesterase with broad substrate specificity causes organophosphorus, carbamate and pyrethroid resistance in peach–potato aphids (*Myzus persicae*). *Pestic. Biochem. Physiol.* **18**, 235–246.

Devonshire, A. L., Moores, G. D. & Chiang, C. 1983 The biochemistry of insecticide resistance in the peach–potato aphid (*Myzus persicae*). *Pesticide Chemistry (Proc. 5th International Congress of Pesticide Chemistry)* **3**, 191–196.

Devonshire, A. L., Byrne, F. J., Moores, G. D. & Williamson, M. S. 1998 Biochemical and molecular characterisation of insecticide insensitive acetylcholinesterase in resistant insects. In *Structure and function of cholinesterases and related proteins* (ed. B. P. Doctor, D. M. Quinn, R. L. Rotundo and P. Taylor). New York: Plenum. (In the press.)

Dewar, A. M., Haylock, L. A., Chapman, J., Devine, G. J., Harling, Z. & Devonshire, A. L. 1994 Effect of triazamate on resistant *Myzus persicae* on sugar beet under field cages. In *Brighton Crop Protection Conference: Pests and Diseases*, pp. 407–412. Bracknell, UK: BCPC.

Eggers-Schumacher, H. A. 1983 A comparison of the reproductive performance of insecticide-resistant and susceptible clones of *Myzus persicae*. *Entomol. Exp. Appl.* **34**, 301–307.

Elbert, A., Overbeck, H., Iwaya, K. & Tsuboi, S. 1990 Imidacloprid, a novel systemic nitromethylene analogue insecticide for crop protection. In *Brighton Crop Protection Conference: Pests and Diseases*, pp. 21–28. Bracknell, UK: BCPC.

ffrench-Constant, R. H., Devonshire, A. L. & Clark, S. J. 1987 Differential rate of selection for resistance by carbamate, organophosphorus and combined pyrethroid and organophosphorus insecticides in *Myzus persicae* (Sulzer) (Hemiptera: Aphididae). *Bull. Entomol. Res.* **77**, 227–238.

ffrench-Constant, R. H., Harrington, R. & Devonshire, A. L. 1988*a* Effect of repeated application of insecticides to potatoes on numbers of *Myzus persicae* (Sulzer) (Hemiptera: Aphididae) and on the frequencies of insecticide resistant variants. *Crop Protection* **7**, 55–61.

ffrench-Constant, R. H., Devonshire, A. L. & White, R. P. 1988*b* Spontaneous loss and reselection of resistance in extremely resistant *Myzus persicae* (Sulzer). *Pestic. Biochem. Physiol.* **30**, 1–10.

Field, L. M. & Devonshire, A. L. 1997 Structure and organisation of amplicons containing the E4 esterase genes responsible for insecticide resistance in the aphid *Myzus persicae* (Sulzer). *Biochem. J.* **322**, 867–871.

Field, L. M. & Devonshire, A. L. 1998 Evidence that the E4 and FE4 esterase genes responsible for insecticide resistance in the aphid *Myzus persicae* (Sulzer) are part of a gene family. *Biochem. J.* **330**, 169–173.

Field, L. M., Devonshire, A. L. & Forde, B. G. 1988 Molecular evidence that insecticide resistance in peach–potato aphids (*Myzus persicae*, (Sulz.)) results from amplification of an esterase gene. *Biochem. J.* **251**, 309–312.

Field, L. M., Devonshire, A. L., ffrench-Constant, R. H. & Forde, B. G. 1989 Changes in DNA methylation are associated with loss of insecticide resistance in the peach–potato aphid *Myzus persicae* (Sulz.). *FEBS Lett.* **243**, 323–327.

Field, L. M., Williamson, M. S., Moores, G. D. & Devonshire, A. L. 1993 Cloning and analysis of the esterase genes, conferring insecticide resistance in the peach–potato aphid *Myzus persicae* (Sulzer). *Biochem. J.* **294**, 569–574.

Field, L. M., Javed, N., Stribley M. F. & Devonshire, A. L. 1994 The peach–potato aphid *Myzus persicae* and the tobacco aphid *Myzus nicotianae* have the same esterase-based mechanisms of insecticide resistance. *Insect Molec. Biol.* **3**, 143–148.

Field, L. M., Devonshire, A. L. & Tyler-Smith, C. 1996 Analysis of amplicons containing the esterase genes responsible for insecticide resistance in the peach–potato aphid *Myzus persicae* (Sulzer). *Biochem. J.* **313**, 543–547.

Field, L. M., Anderson, A. P., Denholm, I., Foster, S. P., Harling, Z. K., Javed, N., Martinez-Torres, D., Moores, G. D., Williamson, M. S. & Devonshire, A. L. 1997 Use of biochemical and DNA diagnostics for characterising multiple mechanisms of insecticide resistance in the peach–potato aphid, *Myzus persicae* (Sulzer). *Pestic. Sci.* **51**, 283–289.

Foster, S. P., Harrington, R., Devonshire, A. L., Denholm, I., Devine, G. J. & Kenward, M. G. 1996 Comparative survival of insecticide-susceptible and resistant peach-potato aphids, *Myzus persicae* (Sulzer) (Hemiptera: Aphididae), in low temperature field trials. *Bull. Entomol. Res.* **86**, 17–27.

Foster, S. P., Harrington, R., Devonshire, A. L., Denholm, I., Clark, S. J. & Mugglestone, M. A. 1997 Evidence for a possible fitness trade-off between insecticide resistance and the low temperature movement that is essential for survival of UK populations of *Myzus persicae* (Hemiptera: Aphididae). *Bull. Entomol. Res.* **87**, 573–579.

Foster, S. P., Denholm, I., Harling, Z. K., Moores, G. D. & Devonshire, A. L. 1998 Intensification of resistance in UK field populations of the peach–potato aphid, *Myzus persicae* (Homoptera: Aphididae) in 1996. *Bull. Entomol. Res.* **88**, 127–130.

Frey, J. E. 1993 The analysis of athropod pest movement through trade in ornamental plants. In *Brighton Crop Protection Conference Monograph no. 54. Plant Health and the European Single Market*, pp. 157–165. Bracknell, UK: BCPC.

Hedley, D., Khambay, B. P. S., Hooper, A. M., Thomas R. D. & Devonshire, A. L. 1998 Proinsecticides effective against insecticide-resistant peach–potato aphids (*Myzus persicae* (Sulzer)). *Pestic. Sci.* **53**, 201–208.

Hick, C. A., Field, L. M. & Devonshire, A. L. 1996 Changes in the methylation of amplified esterase DNA during loss and reselection of insecticide resistance in peach–potato aphids, *Myzus persicae*. *Insect. Biochem. Molec. Biol.* **26**, 41–47.

Holliday, R., Ho, T. & Paulin, R. 1996 Gene silencing in mammalian cells. In *Epigenetic mechanisms of gene regulation*, pp. 47–59. New York: Cold Spring Harbor Laboratory.

Lilly, M., Kreber, R., Ganetsky, B. & Carlson, J. R. 1994 Evidence that the *Drosophila* olfactory mutant *smellblind* defines a novel class of sodium channel mutation. *Genetics* **136**, 1087–1096.

Martinez-Torres, D., Devonshire, A. L. & Williamson, M. S. 1997 Molecular studies of knockdown resistance to pyrethroids: cloning of domain II sodium channel gene sequences from insects. *Pestic. Sci.* **51**, 265–270.

Martinez-Torres, D., Foster, S. P., Field, L. M., Williamson, M. S. & Devonshire, A. L. 1998 A sodium channel mutation is associated with knockdown resistance (kdr) to DDT and pyrethroids in the peach–potato aphid, *Myzus persicae* (Sulzer) (Hemiptera: Aphididae). *Insect Molec. Biol.* (In the press.)

Miyazaki, M., Ohyama, K., Dunlap, D. Y. & Matsumura, F. 1996 Cloning and sequencing of the *para*-type sodium channel gene from susceptible and *kdr*-resistant German cockroaches (*Blattella germanica*) and housefly (*Musca domestica*). *Molec. Gen. Genet.* **252**, 61–68.

Moores, G. D., Devine, G. J. & Devonshire, A. L., 1994*a* Insecticide-insensitive acetylcholinesterase can enhance esterase-based resistance in *Myzus persicae* and *Myzus nicotianae*. *Pestic. Biochem. Physiol.* **49**, 114–120.

Moores, G. D., Devine, G. J. & Devonshire, A. L. 1994*b* Insecticide resistance due to insensitivie acetylcholinesterase in *Myzus persicae* and *Myzus nicotianae*. In *Brighton Crop Protection Conference: Pests and Diseases*, pp. 413–418. Bracknell, UK: BCPC.

Muggleton, J., Hockland, S., Thind, B. B., Lane, A. & Devonshire, A. L. 1996 Long-term stability in the frequency of insecticide resistance in the peach–potato aphid, *Myzus persicae*, in England. In *Brighton Crop Protection Conference: Pests and Diseases*, vol. 2, pp. 739–744. Bracknell, UK: BCPC.

Mutero, A., Pralavorio, M., Bride, J.-M & Fournier, D. 1994 Resistance-associated point mutations in insecticide-insensitive acetylcholinesterase. *Proc. Natn. Acad. Sci. USA* **91**, 5922–5926.

Needham, P. H. & Sawicki, R. M. 1971 Diagnosis of resistance to organophosphorus insecticides in *Myzus persicae* (Sulz.). *Nature* **230**, 125–126.

Pasteur, N. & Raymond, M. 1996 Insecticide resistance genes in mosquitoes: their mutations, migration, and selection in field populations. *J. Heredity* **87**, 444–449.

Sawicki, R. M., Devonshire, A. L., Payne, R. W. & Petzing, S. M. 1980 Stability of insecticide resistance in the peach–potato aphid, *Myzus persicae* (Sulzer). *Pestic. Sci.* **11**, 33–42.

Smith, S. D. J., Dewar, A. M. & Devonshire, A. L. 1990 Resistance of *Myzus persicae* to insecticides applied to sugar beet. In *IIRB 53rd Winter Congress*, pp. 379–398.

Stark, G. R. 1993 Regulation and mechanisms of mammalian gene amplification. In *Gene amplification in mammalian cells* (ed. R. E. Hellems), pp. 243–254. New York: Marcel Dekker.

Williamson, M. S., Martinez-Torres, D., Hick, C. A. & Devonshire, A. L. 1996 Identification of mutations in the housefly *para*-type sodium channel gene associated with knockdown resistance (*kdr*) to pyrethroid insecticides. *Molec. Gen. Genet.* **252**, 51–60.

Why are there so few resistance-associated mutations in insecticide target genes?

Richard H. ffrench-Constant, Barry Pittendrigh, Ashley Vaughan and Nicola Anthony

Department of Entomology and Center for Neuroscience, 237 Russell Laboratories, 1630 Linden Drive, University of Wisconsin-Madison, Madison, WI 53706, USA (ffrench@vms2.macc.wisc.edu)

The genes encoding the three major targets of conventional insecticides are: *Rdl*, which encodes a γ-aminobutyric acid receptor subunit (RDL); *para*, which encodes a voltage-gated sodium channel (PARA); and *Ace*, which encodes insect acetylcholinesterase (AChE). Interestingly, despite the complexity of the encoded receptors or enzymes, very few amino acid residues are replaced in different resistant insects: one within RDL, two within PARA and three or more within AChE. Here we examine the possible reasons underlying this extreme conservation by looking at the aspects of receptor and/or enzyme function that may constrain replacements to such a limited number of residues.

Keywords: acetylcholinesterase; *Rdl*; γ-aminobutyric acid receptor; *para*; voltage-gated sodium channel; insecticide resistance

1. INTRODUCTION

The three major targets of conventional insecticides are: (i) the γ-aminobutyric acid (GABA) receptor containing RDL subunits encoded by the gene *Resistance to dieldrin* or *Rdl*, the target for cyclodiene insecticides and the recently introduced fipronils; (ii) the PARA voltage-gated sodium channel encoded by the gene *para*, the target site for DDT and pyrethroids; and (iii) insect acetylcholinesterase (AChE) encoded by the gene *Ace*, the target site for organophosphorus (OP) and carbamate insecticides. Genes encoding all three of these target sites have been cloned from *Drosophila melanogaster*, supporting the previously proposed role of this insect as a model in which to study insecticide resistance (Wilson 1988; ffrench-Constant *et al.* 1992). However, homologues of these *Drosophila* target-site genes have recently also been cloned from a range of pest insects and the underlying resistance-associated mutations compared. Interestingly, despite the wide range of insects studied, and presumed differences in the modes of insecticide selection, the same residues are consistently replaced in the same receptors and/or enzymes: a single residue in RDL, two residues in PARA and three or more in AChE.

The purpose of the current paper is to examine the question: why are there so few resistance-associated mutations in the genes encoding these receptors and/or enzymes? This will be achieved by examining our work on receptor and/or enzyme function in relation to the likely functional constraints placed on these important components of the insect nervous system. In each case a working model or hypothesis is presented to account for the conservation of amino-acid replacements observed in the context of what we know about the normal function of the receptor and which residues may interact with the insecticide.

2. CONSERVATION OF RESISTANCE-ASSOCIATED MUTATIONS

(a) *The* **Rdl**-*encoded GABA receptor*

The gene *Rdl*, encoding the GABA receptor subunit RDL, was cloned from a *D. melanogaster* mutant resistant to cyclodiene insecticides and picrotoxinin (PTX), a vertebrate GABA$_A$ receptor antagonist (ffrench-Constant *et al.* 1991). The pharmacology of RDL-containing insect receptors and the relationship of these GABA receptors to those found in vertebrates has recently been reviewed elsewhere (Hosie *et al.* 1997). To determine the resistance-associated mutation(s), we examined worldwide collections of resistant *D. melanogaster*. In these strains we consistently documented the replacement of the same amino acid, alanine302 with serine (ffrench-Constant *et al.* 1993b). Following functional expression of RDL subunits as homomultimers in a range of heterologous expression systems (ffrench-Constant *et al.* 1993a; Lee *et al.* 1993), we also confirmed the functional relevance of this mutation by showing that replacement of alanine302 with serine results in insensitivity of the resulting GABA-gated chloride channels (ffrench-Constant *et al.* 1993a). Importantly, RDL-containing GABA receptor subunits are widely expressed in the insect nervous system (Aronstein & ffrench-Constant 1995; Aronstein *et al.* 1996); despite the fact that we can document that RDL co-assembles with another unidentified subunit (via observed differences of channel conductance *in vivo* and *in vitro*) (Zhang *et al.* 1995), RDL expression *in vitro* reconstitutes most of the pharmacology observed in insect GABA receptors *in vivo* (ffrench-Constant *et al.* 1993a; Zhang *et al.* 1994, 1995). A survey of a wide range of other insect species, including several beetles, a mosquito (*Aedes aegypti*), the whitefly *Bemisia tabaci*, and a cockroach (*Blattella germanica*) (Thompson *et al.* 1993; ffrench-Constant *et al.* 1994;

Anthony *et al.* 1995*b*), showed that this alanine-to-serine replacement was the most common resistance-associated replacement. More rarely, in a different fruit fly, *D. simulans* (ffrench-Constant *et al.* 1993*b*), and in the aphid, *Myzus persicae* (N. Anthony and R. ffrench-Constant, unpublished data), the same residue is replaced with a glycine.

(b) *The PARA voltage-gated sodium channel*

The gene *para*, which encodes the PARA voltage-gated sodium channel, was cloned from *Drosophila* in the laboratory of Barry Ganetzky on the basis of the *temperature sensitive* paralytic phenotypes displayed by *para^{ls}* alleles (Loughney *et al.* 1989). The PARA sodium channel appears to be the major sodium channel in insects and is a large polypeptide composed of four homology domains (I–IV), each containing six proposed hydrophobic membrane-spanning domains (S1–S6) (Loughney *et al.* 1989). After linkage studies that correlated resistance to DDT and pyrethroids with the location of the *para*-homologous sodium-channel gene in *knockdown resistant (kdr)* house flies (Williamson *et al.* 1993; Knipple *et al.* 1994) and cockroaches (Dong & Scott 1994), *para* homologues were cloned from both species and the underlying resistance-associated mutations examined (Miyazaki *et al.* 1996; Williamson *et al.* 1996; Dong 1997). Again, amino-acid replacements were confined to only two positions. The first, associated with the original *kdr* strain, was in the S6 hydrophobic segment of homology domain II (termed IIS6). The second, associated with another more resistant allele, termed *super-kdr*, was found as a double mutant with both the *kdr* mutation and a second replacement in the intracellular loop between IIS4 and IIS5. The '*kdr*-like' replacement is similar in the housefly (Miyazaki *et al.* 1996; Williamson *et al.* 1996), the horn fly (Guerrero *et al.* 1997), the cockroach (Dong 1997), the tobacco budworm *Heliothis virescens* (Park & Taylor 1997), the aphid *M. persicae* (Field *et al.* 1997), and in the mosquito *Anopheles gambiae* (Martinez-Torres *et al.* 1998), whereas the second *super-kdr* type mutation has only been found in resistant houseflies (Williamson *et al.* 1996) and horn flies (Guerrero *et al.* 1997). Interestingly, so far these replacements associated with field-collected resistance alleles have been found to reside within the second homology domain of the protein.

Recently, the housefly *para* homologue has been functionally expressed in *Xenopus* oocytes and the leucine1014 to phenylalanine replacement (*kdr* mutation) has been shown to confer ten-fold insensitivity to cismethrin on the resulting sodium channels (Smith *et al.* 1997).

(c) *The acetylcholinesterase-encoding locus,* Ace

AChE degrades the neurotransmitter acetylcholine in the insect synapse and is the primary target site for OP and carbamate insecticides, which inhibit the enzyme. The gene *Ace*, which encodes insect acetylcholinesterase (AChE), was again cloned from *Drosophila* based on the knowledge of its location via isolation of a range of different mutants (Hall & Spierer 1986). The screening of a range of different *D. melanogaster* and housefly strains has identified a range of different amino-acid replacements putatively causing resistance (reviewed by Feyereisen (1995)). The functional significance of several of these have been tested in the *Drosophila* enzyme following

the heterologous expression of different mutants in *Xenopus* oocytes and direct testing of their insensitivity to a range of different insecticide inhibitors (Mutero *et al.* 1994). A comparison of the different putative resistance-associated amino-acid replacements in the fruit fly and the housefly (Feyereisen 1995) again suggests that only a limited subset of replacements cause resistance despite the widely differing sizes and structures of different OP and carbamate insecticides. However, many of these resistance-associated residues are predicted to lie within or close to the active-site gorge of this enzyme, based on superimposition of the insect amino-acid sequence on the three-dimensional crystal structure obtained from *Torpedo* (Mutero *et al.* 1994). The role of other replacements linked to AChE insensitivity in other insects, such as the Colorado potato beetle *Leptinotarsa decemlineata* (Zhu & Clark 1997), remains to be functionally proven.

3. IMPLICATIONS FOR RECEPTOR AND ENZYME STRUCTURE AND FUNCTION

(a) *The RDL containing GABA receptor*

RDL GABA receptor subunits co-assemble in a variety of expression systems to give functional homomultimeric GABA-gated chloride-ion channels (ffrench-Constant *et al.* 1993*a*; Lee *et al.* 1993). Further, despite the fact that the conductance of these channels differs from those of RDL-containing channels in the insect nervous system (Zhang *et al.* 1994, 1995) (showing that other subunits are present in RDL-containing receptors in the insect nervous system), these homomultimers reconstitute much of the GABA receptor pharmacology displayed *in vivo*, notably PTX and cyclodiene sensitivity and bicuculline insensitivity (Zhang *et al.* 1995). The observation that the replacement of alanine302 with serine confers insecticide insensitivity leads to the simplest hypothesis, that PTX and cyclodienes actually bind within the ion-channel pore (as alanine302 is within the second membrane-spanning region, which is predicted to line the chloride ion-channel pore by homology with structural and modelling work performed in the closely related nicotinic acetylcholine receptor (Leonard *et al.* 1988)). However, our consistent finding of replacements of this same amino acid (with either a serine or a glycine) in a wide range of different insects leads us to examine why only this residue is replaced in resistance. Thus, for example, the binding site of PTX and cyclodienes is unlikely to interact with only this single residue, therefore why can other residues in the binding site not be replaced to give resistance?

To examine this question we performed a biophysical analysis of wild-type and mutant RDL receptors via the patch clamping of cultured *Drosophila* neurons (Zhang *et al.* 1994). Detailed analysis revealed significant but small shifts in a range of channel parameters, namely small changes in the shape of the GABA dose–response curves, small changes in both the inward and the outward conductance of the channel, and a net stabilization of the channel's open state. However, the largest change in a single channel parameter was in the rate of receptor desensitization (Zhang *et al.* 1994; ffrench-Constant *et al.* 1995). This can be described as the rate at which the channel returns to its normal activatable state under prolonged exposure to GABA (figure 1*a*). Thus, channels

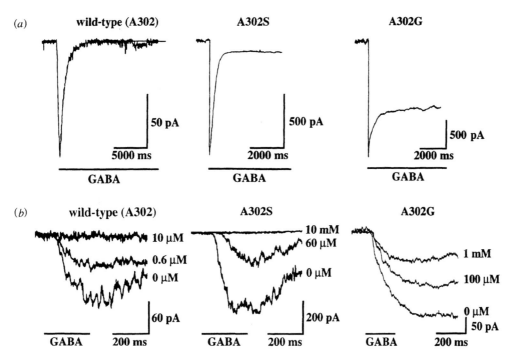

Figure 1. Decreased rates of GABA receptor desensitization are correlated with increased levels of PTX resistance in two different *Rdl* mutants. (*a*) Rates of receptor densitization in wild-type (alanine302) neurons and in those from two different resistant mutants: alanine302 to serine (A302S) and alanine302 to glycine (A302G). (*b*) Levels of PTX resistance observed in the same mutants. Note that wild-type receptors can be blocked by 10 μM PTX whereas receptors containing alanine302 to serine require 10 mM and those containing alanine302 to glycine also require more than 1 mM for partial block.

containing alanine302 to serine replacements desensitized 29 times more slowly than wild-type ones and desensitization was incomplete (figure 1*a*) Although at first these changes are difficult to reconcile with the interaction of the receptor with the antagonist, other workers have proposed that PTX binds preferentially to the desensitized state of the vertebrate GABA$_A$ receptor (Newland & Cull-Candy 1992). We have therefore formulated a working model whereby replacements of alanine302 have a dual role in both: (i) interacting directly with the drug binding site within the ion-channel pore, and (ii) allosterically destabilizing the insecticide's preferred desensitized state of the RDL-containing GABA receptor (Zhang *et al.* 1994; ffrench-Constant *et al.* 1995).

Our recent experiments on cultured *Drosophila* neurons have added weight to this hypothesis by demonstrating that the replacement of alanine302 with a glycine residue (the alternative resistance-associated replacement found in nature) also confers resistance to PTX and is again (like alanine302 to serine) associated with slower and incomplete receptor densensitization (compare figure 1*a* and 1*b*). Further experiments have been performed to test a corollary of this hypothesis: that replacements of other residues known to affect rates of receptor desensitization should also affect resistance but to a lesser extent. In conclusion, current data are consistent with a model whereby replacements of alanine302 cause resistance by interacting directly with the insecticide binding site and also allosterically by destabilizing the insecticide's preferred (desensitized) conformation of the receptor. This implies that replacements having effects on only one process or the other would not confer sufficient resistance on the insect for insecticide selection to be effective in the field.

(b) *The PARA voltage-gated sodium channel*

Owing to the relative ease of isolating large numbers of *para* mutants based on their *temperature sensitive* (*ts*) phenotypes, we screened an existing collection of *ts* mutants to indentify any that were also resistant to DDT and pyrethroid insecticides (Pittendrigh *et al.* 1997). This approach has the advantage of using temperature sensitivity as an independent screen for replacements that affect insecticide sensitivity of the channel, and may therefore identify residues that interact with insecticides but may carry sufficient fitness costs in the whole insect to prevent their selection in natural populations. This technique can therefore potentially identify a larger range of residues affecting the interaction of the channel with insecticides than those found in nature.

We examined DDT resistance in a group of 12 independent *para* mutants, including two that cause dominant temperature-sensitive paralysis, nine showing recessive temperature sensitivity and one that causes a smell-blind phenotype (Pittendrigh *et al.* 1997). Of these 12 mutants, we found that six were also associated with DDT insensitivity. As expected, resistance was sex-linked, as *para* is X-linked in *D. melanogaster*, and not responsive to the synergist piperonyl butoxide (PBO), which suppresses metabolic resistance (Pittendrigh *et al.* 1997). Two general aspects of the overall nature of these mutations are noteworthy. First, exactly half of the temperature-sensitive mutations examined conferred resistance, despite their not being preselected with insecticides. This observation may imply that only a limited number of replacements can be generated in the sodium channel polypeptide which confer a temperature-sensitive and/or resistant phenotype. Second, although many of these mutations occupy positions equivalent to the *kdr* and *super-kdr* mutations (figure 2), none reside in domain II. This might imply that replacements in homology domain II are less prone to confer temperature-sensitive phenotypes and this avoidance of a potentially adverse fitness cost may be part of the reason why most resistance-associated replacements appear confined to this domain in natural populations. However, alternatively, we note that a further 'kdr-like' replacement (i.e. located in the equivalent section of S6) correlated with resistance in *Heliothis virescens* lies in homology domain I and not domain II (Park & Taylor 1997). This observation suggests that resistance-associated replacements can be found in other homology domains and that their previous documentation solely in domain II may just be an artefact of (i) the limited number of species examined, and (ii) the emphasis on re-examining the same region, i.e. II S6.

Several specific aspects of the relative locations of the different replacements in *D. melanogaster* are also interesting (Pittendrigh *et al.* 1997). Thus the replacements fall into three categories (figure 2): (i) those that are 'kdr-like', such as *para^{74}*, which resides in an equivalent position in S6 but in the third rather than the second homology domain; (ii) those that are 'super-kdr-like', such as *para^{DN7}* and *para^{ts1}*/*para^{ts2}* (independent occurrences of the same point mutation), which lie within the intracellular S4–S5 linker but in domains III and I, respectively; and (iii) those that occupy a novel position, such as *para^{DTS2}*/*para^{DN43}* (again independent examples of the same mutation), which lie within the S5–S6 loop that may actually form part of the sodium-channel lining. In terms of resistance levels generated, independently, the 'kdr-like' or 'super-kdr-like' replacements in *D. melanogaster para* both confer low levels of resistance to DDT. Interestingly, and similarly to *super-kdr* in the housefly, combination of a 'kdr-like' and 'super-kdr-like' mutation in *Drosophila* heterozygous for the two alleles also gives increased resistance to the type II (cyano-group-containing) pyrethroids in the whole fly.

Again, the simplest interpretation of these data is that these replacements all denote the binding site for either DDT and type I ('kdr-like' replacements) or additionally type II ('super-kdr-like' replacements) pyrethroids (Williamson *et al.* 1996). However, our results suggest the possibility of a different mechanism. The identification of DDT-resistant mutations at sites nearly equivalent to *kdr* and *super-kdr* but in other homology domains suggests that if these mutations do indeed define a binding site, it is likely to be composed of residues contributed by each of the four homology domains. This possibility seems reasonable given the essentially tetrameric structure of the sodium-channel α-subunit. However, it is also possible that the sodium-channel mutations characterized here, and in previous studies, confer resistance to pyrethroids and DDT by a mechanism other than the direct alteration of the binding site. The pharmacological effect of these insecticides is to cause persistent activation of sodium channels by delaying the normal voltage-dependent mechanism of inactivation (Soderlund & Bloomquist 1989). Rather than directly affecting the binding of pyrethroids and DDT, the mutations may confer resistance by causing functional changes in sodium-channel properties that compensate for or that alleviate the consequences of

the insecticide. For example, if the mutations altered the voltage dependence or kinetics of activation or inactivation, the neurotoxic effects of the insecticide could be reduced or overridden. In this context we note that recent functional analysis of the equivalent of the *kdr*-associated amino-acid replacement in a sodium channel in the rat has provided evidence for accompanying changes in voltage dependence (Vais *et al.* 1997).

This interpretation is particularly consistent with the resistant mutations that map to the S4–S5 loop in the different homology domains, because a variety of evidence from various systems suggests the importance of this region in channel function. Mutations in this loop are known to reduce fast inactivation of potassium channels (Isacoff *et al.* 1991). Furthermore, mutations in mammalian sodium channels at or near the site defined by the *para^{DN7}* mutation are associated with a variety of abnormalities. One form of long QT syndrome, an inherited cardiac arrhythmia, is caused by an amino-acid substitution in the SCN5A channel at a site adjacent to the residue affected by *para^{DN7}* (Wang *et al.* 1995). At the same residue as *para^{DN7}*, an alanine-to-threonine replacement in the human SCN4A channel causes paramyotonia congenita, a disorder associated with decreased kinetics of sodium-channel inactivation (McClatchey *et al.* 1992; Yang *et al.* 1994). The same replacement in a mouse neuronal sodium channel produces the *jolting* phenotype (Kohrman *et al.* 1996). Functional studies indicate that this mutation significantly shifts the voltage dependence of activation in the depolarizing direction. The *para^{ts1}* mutation resides at the identical location in homology domain I and may cause functional perturbations analogous to those caused by mutations in the vicinity of *para^{DN7}*. The *kdr*/*para^{74}* mutations in S6 transmembrane segments might also alter the gating properties of sodium channels in some distinctive manner.

Our finding of synergistic resistance to pyrethroids in *para^{74}*/*para^{DN7}* heterozygotes is of particular interest. The changes conferred by these two mutations in homology domain III are in locations analogous to those of the two mutant sites in homology domain II in *super-kdr* strains of houseflies; the elevated levels of resistance in the heterozygote approximate the enhanced resistance in *kdr* compared with *super-kdr* strains. However, the two circumstances are very different. In *super-kdr* mutants, the two lesions reside in the same polypeptide (i.e. they are in *cis*); in *para^{74}*/*para^{DN7}* heterozygotes, the lesions reside in different polypeptides (i.e. in *trans*). The increased resistance in *super-kdr* housefly strains is presumed to reflect a combined deficit in pyrethroid binding caused by the doubly mutant polypeptide. This explanation seems inadequate to account for the phenotype of *para^{74}*/*para^{DN7}* heterozygotes. However, this phenotype can be reconciled with the alternative view that resistance is mediated via functional alterations of the encoded sodium-channel polypeptides. The particular functional defect associated with one mutation could enhance the perturbation caused by the other. For example, one mutation could alter the inactivation mechanism, leading to a slight depolarization of membrane potential and the other could alter the voltage dependence of inactivation. The combined effect in double heterozygotes could be very different from the effect of either mutation alone or when heterozygous with

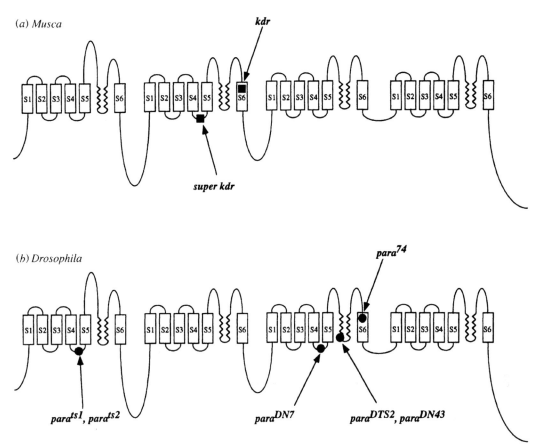

Figure 2. Comparison of the relative locations of DDT/pyrethroid resistance-associated replacements in the *para*-encoded voltage-gated sodium channel of *Drosophila melanogaster* and those in the house fly *Musca domestica*. (*a*) The location of the replacements associated with *kdr* and *super-kdr* in the *para*-homologous channel from *Musca*. Note that these are both in homology domain II. (*b*) The location of replacements in *Drosophila para*, which were isolated on the basis of temperature-sensitive paralytic phenotypes but which also confer DDT/pyrethroid resistance. Two classes of these mutations are in positions approximately equivalent to those in *kdr* and *super-kdr*; the position replaced in both *para*[DTS2] and *para*[DN43] represents a novel third class of resistance-associated mutation. Note, however, that none of the temperature-sensitive replacements resides in domain II.

a wild-type allele. This interpretation is also consistent with the observed phenotypic dominance of the double heterozygote in the presence of an extra copy of the *para* locus encoding wild-type sodium-channel subunits. This situation is analogous to the dominant effect of mutations causing hyperkalaemic periodic paralysis (Ptacek *et al.* 1991; Rojas *et al.* 1991; McClatchey *et al.* 1992) in humans, where the misbehaviour of a small percentage of mutant sodium channels at the resting membrane potential shifts all of the other channels (including wild-type channels) to an inactivated state (Cannon *et al.* 1991).

We are currently performing a direct electrophysiological analysis of the resistant *Drosophila para* mutants to confirm their importance in defining resistance-associated replacements and also to separate the relative roles of the different mutations in altering channel function and/or insecticide binding. To date, our results support a hypothesis whereby some replacements interact with channel inactivation. Thus, for example, the application of 500 nM allethrin to wild-type channels removed inactivation and prolonged tail currents whereas it had little or no effect on *para*[74] mutant channels (R. Martin, B. Pittendrigh, J. Liu, R. Reenan, B. Ganetzky and D. Hanck, unpublished data). However, we will need to examine '*kdr*-like' and '*super-kdr*-like' mutations independently and

together in the same fly (in *trans* rather than in *cis*) to test any potential dual effect on channel function and insecticide insensitivity.

(c) *Acetylcholinesterase*

To date, the functional relevance of potential resistance-associated replacements in insect AChE has only been tested in *D. melanogaster* (Mutero *et al.* 1994). Therefore, to confirm the relevance of putative resistance-associated mutations in a pest insect we have chosen the *Ace* gene from the yellow fever mosquito, *A. aegypti*, as a model system in which to study the effects of mutations on a pest insect enzyme (Anthony *et al.* 1995a; Vaughan *et al.* 1997). This mosquito is a vector of both yellow fever and dengue; we have recently found insensitive AChE in populations of this species in Trinidad (Vaughan *et al.* 1998).

Acetylcholinesterase genes (*Ace* homologues) have now also been cloned from a number of insect species including the mosquitoes *Anopheles stephensi* (Hall & Malcolm 1991) and *Culex pipiens* (C. Malcolm, unpublished data). However, although insecticide-insensitive AChE has been widely documented biochemically from a range of both *Culex* and *Anopheles* mosquitoes (Ayad & Georghiou 1975; Hemingway & Georghiou 1983; Hemingway *et al.* 1985; Raymond *et al.* 1986; Haas *et al.*

Table 1. *Residues replaced in* Aedes aegypti *AChE compared with those found in other insects with insensitive AChE*

(Only the replacements in *Drosophila* have been tested by site-directed mutagenesis and functional expression. Note that the mutations made in the *Aedes* enzyme correspond to those found in other insects and not to any yet found in natural populations of *A. aegypti.*)

| insect | strain | numbering in AChE of *Torpedo californica* | | | | | | |
		78	129	151	227	238	288	328
Drosophila melanogaster	Saltillo	Phe > Ser	Ile > Val	—	Gly > Ala	—	Phe > Tyr	—
	Bygdea	—	Ile > Val	—	Gly > Ala	—	—	—
	Pierrefeu	—	Ile > Thr	—	Gly > Ala	—	—	—
	MH19	—	—	—	—	—	Phe > Tyr	—
Musca domestica	77M	—	—	Val > Leu	Gly > Ala	—	Phe > Tyr	—
	CH2	—	—	—	Gly > Ala	—	Phe > Tyr	—
	49R	—	—	—	—	—	—	Gly > Ala
Leptinotarsa decemlineata	AZ-R	—	—	—	—	Ser > Gly	—	—
mutagenized *Aedes aegypti*	—	Phe > Ser	—	—	Gly > Ala	—	Phe > Tyr	—

1988; ffrench-Constant & Bonning 1989; Tang & Cammak 1990), no resistance-associated mutations have been described from mosquitoes themselves.

Analysis of insecticide-insensitive AChE in mosquitoes has also recently been complicated by the discovery of two AChEs (AChE1 and AChE2) in *Culex pipiens* (Bourguet *et al.* 1996*b*). These differ in their sensitivity to insecticide inhibition. In susceptible insects, AChE1 can be inhibited by a fixed dose of a carbamate insecticide (5×10^{-1} M propoxur) whereas AChE2 is unaffected by this concentration (Bourguet *et al.* 1996*a*). Linkage mapping of the *Ace* gene of *Culex pipiens* suggests that it is at a sex-linked locus (C. Malcolm, personal communication), whereas resistance in *Culex* is not sex-linked. However, detailed analysis of inhibition profiles in a range of other mosquito species, including *Aedes aegypti*, suggests that most other mosquitoes bear only a single *Ace* locus and that *C. pipiens* may therefore be an exception. Interestingly, linkage mapping of the *A. aegypti* clone discussed here shows that it maps extremely close to the sex-determining locus (Severson *et al.* 1997); this result suggests that resistance should be sex-linked.

We cloned a section of the *Ace* locus from *A. aegypti* with the use of degenerate primers in the polymerase chain reaction and then isolated a full-length clone from an adult cDNA library (Anthony *et al.* 1995*a*). Functional expression was achieved in baculovirus-infected SF21 insect cells (Anthony *et al.* 1995*a*) and the effects of the resistance-associated amino-acid replacements found in *Drosophila* and the housefly were tested by mutagenesis of the *Aedes* gene *in vitro* (table 1). Resistance of AChE to OP and carbamate insecticides is due to modifications of the active site (amino-acid replacements) and these modifications also appear to alter the catalytic activity of the enzyme towards insecticides and substrates. Therefore, the mutant forms of *Aedes* AChE all behaved differently from the wild-type when initial rates of activity were assayed with acetyl, propionyl and butyryl thiocholine iodides (ASCI, PSCI and BSCI). In most cases, the initial rate was highest with ASCI, followed by PSCI and then BSCI. However, in the single mutant F350Y and the double mutant F105S+F350Y, higher rates of activity with PSCI and

BSCI than with ASCI were observed, for which the activity was approximately 20% of that of the wild-type enzyme (Vaughan *et al.* 1997). In *D. melanogaster*, the corresponding F368Y mutant also had higher rates of activity with PSCI and BSCI (Fournier *et al.* 1996). Furthermore, decreased rates of reaction with ASCI for insensitive forms of AChE have been documented in both *Anopheles albimanus* and *C. pipiens* (ffrench-Constant & Bonning 1989; Bourguet *et al.* 1996*a*). It will therefore be interesting to investigate, when rates of reaction with ASCI fall, whether replacements equivalent to F368Y are found.

In terms of resistance, mutagenesis of *D. melanogaster Ace* showed a correlation between the number of mutations in the expressed enzyme and the bimolecular constant (k_i) ratio of the mutated compared with the wild-type from of the enzyme (Mutero *et al.* 1994). In general, the greater the number of resistance-associated amino-acid replacements, the higher the resistance ratio, although this was not always the case. In the mutagenized *Aedes* enzyme, a similar pattern of relative insensitivity was found (figure 3). However, the comparison is complicated by the presence of one of the mutations in the wild-type form of the *A. aegypti Ace*: the Ile199Val mutation in *D. melanogaster* AChE is already present in *A. aegypti* AChE (Val185). Whereas in *D. melanogaster* none of the single point mutations gave rise to significant levels of resistance, the G285A mutant gives more than a 20-fold increase in the resistance ratio with the OP paraoxon. For both classes of insecticides, the double mutant G285Y+F350Y and the triple mutant gave substantial increases in the resistance ratio with all the insecticides tested (figure 3).

The three-dimensional structure of *Torpedo californica* AChE has been determined (Sussman *et al.* 1991), and it is thus possible to superimpose other AChEs on this. G303 in *D. melanogaster* (G285 in *A. aegypti*) is thought to affect the orientation of the active-site serine, which is phosphorylated by OPs and carbamylated by carbamates; F368 (F350 in *A. aegypti*) is near the acyl moiety of the bound substrate (Mutero *et al.* 1994). The presence of both these mutations in *A. aegypti* AChE has a profound effect on the binding of insecticide, resulting in the high insensitivity ratios observed.

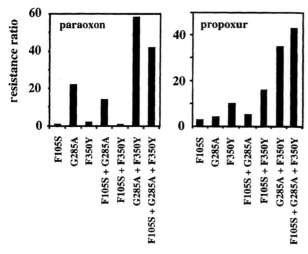

Figure 3. Site-directed mutagenesis of the mosquito *Ace* gene confers insecticide insensitivity on baculovirus-expressed AChE mutant enzymes. Levels of AChE insensitivity to inhibition by paraoxon (an OP) and propoxur (a carbamate). Histograms show the effects on the bimolecular velocity constant (k_i) of single and combined replacements (single-letter code). The ordinate corresponds to the ratio of the resistant to the susceptible k_i, here termed the insensitivity ratio.

These studies have demonstrated that the same mutations that cause insecticide resistance in *D. melanogaster* AChE can also confer insensitivity to the enzyme of the pest mosquito *A. aegypti* when its gene is mutagenized. Given the similarity of the predicted amino-acid sequence of the *Aedes* enzyme with the resistance-associated enzyme from *Drosophila*, the present study forms a useful model for examination of mosquito *Ace* homologues. The ease of expression of the *A. aegypti* AChE enzymes in the baculovirus system should permit a thorough kinetic analysis of enzyme inhibition similar to that performed for the *Drosophila* enzyme. This should give an explanation of how the amino-acid replacements within the resistance-associated AChE mutants affect the binding of insecticide to the enzyme and provide a detailed understanding of their affects on substrate specificity and target-site resistance.

4. CONCLUSIONS

Despite the potential complexity (e.g. the predicted multisubunit nature of RDL-containing GABA receptors), and/or large size (e.g. the PARA voltage-gated sodium channel) of the receptors and enzymes that constitute the primary targets for conventional insecticides, the resistance-associated amino-acid replacements documented to date are strikingly few. In this paper, we have attempted to reconcile this extreme conservation with working hypotheses on the role of these replacements in receptor and/or enzyme insecticide insensitivity. This was achieved by detailed functional analysis of the insect GABA receptor and also by a mutagenesis of the insect voltage-gated sodium channel that relies on a phenotype independent of resistance, namely temperature-sensitive paralysis. This latter mutagenesis underscores the potential importance of temperature-sensitive phenotypes in isolating toxicologically relevant mutations in insects and

in this respect it is also interesting to note with hindsight that *Rdl*[alanine 302 > serine], originally isolated on the basis of insecticide resistance, is also itself a temperature-sensitive paralytic mutant (ffrench-Constant *et al.* 1993*c*).

To explain this striking degree of conservation in replacements in both the GABA-gated chloride channel and potentially the voltage-gated sodium channel, we suggest that the interaction of more than one channel function with insecticide binding sites may be important in constraining the location of resistance-associated replacements. Thus, in the RDL-containing GABA receptor, alanine302 interacts both directly with the insecticide binding site and also allosterically by destabilizing the insecticide's preferred desensitized state of the receptor. In the *para* voltage-gated sodium channel, the observation that individual mutations in different channel polypeptides can combine in *Drosophila* to cause effects similar to those of mutations found in the same polypeptide, also suggests that each mutation may have a unique role in affecting channel function. For example, one replacement could alter the inactivation mechanism, leading to a slight depolarization of the membrane, and the second could alter the voltage dependence of inactivation. In AChE we need to test the functional relevance of other resistance-associated mutations found in pest insects to further test the hypotheses advanced for replacement insecticide interactions in the *Drosophila* enzyme (Mutero *et al.* 1994).

Such studies of the detailed biophysics and biochemistry of the receptors and enzymes associated with insecticide targets not only elucidate potential binding sites but also illustrate that these proteins are dynamic molecules that interact in various conformations with their antagonists and agonists. The effect of resistance-associated replacements may therefore not be easily mimicked by static models based on simple 'lock-and-key' type binding interactions. Although currently confined to the three historically important targets, such studies will also become important in the face of likely insensitivity in new targets such as the nicotinic acetylcholine receptor (nACh), the target for important new compounds such as immidacloprid and spinosad. In the latter case, the close structural relationship between GABA receptors and nACh receptors may forewarn us of potential resistance mechanisms to these highly effective compounds.

I thank all in the ffrench-Constant laboratory past and present that have contributed to this work, R. Roush, in whose laboratory the *Rdl* gene was initially cloned, and also those in the laboratories of our collaborators M. Jackson, D. Hanck and B. Ganetzky.

REFERENCES

Anthony, N., Rocheleau, T., Mocelin, G., Lee, H.-J. & ffrench-Constant, R. H. 1995*a* Cloning, sequencing and functional expression of an acetylcholinesterase gene from the yellow fever mosquito *Aedes aegypti*. *FEBS Lett.* **368**, 461–465.

Anthony, N. M., Brown, J. K., Markham, P. G. & ffrench-Constant, R. H. 1995*b* Molecular analysis of cyclodiene resistance-associated mutations among populations of the sweetpotato whitefly *Bemisia tabaci*. *Pestic. Biochem. Physiol.* **51**, 220–228.

Aronstein, K. & ffrench-Constant, R. 1995 Immunocytochemistry of a novel GABA receptor subunit *Rdl* in *Drosophila melanogaster*. *Invert. Neurosci.* **1**, 25–31.

Aronstein, K., Rocheleau, T. & ffrench-Constant, R. H. 1996 Distribution of two GABA receptor-like subunits in the *Drosophila* CNS. *Invert. Neurosci.* **2**, 115–120.

Ayad, H. & Georghiou, G. P. 1975 Resistance to organophosphates and carbamates in *Anopheles albimanus* based on reduced sensitivity of acetylcholinesterase. *J. Econ. Entomol.* **68**, 295–297.

Bourguet, D., Pasteur, N., Bisset, J. & Raymond, M. 1996*a* Determination of *Ace.1* genotypes in single mosquitoes: toward an ecumenical biochemical test. *Pestic. Biochem. Physiol.* **55**, 122–128.

Bourguet, D., Raymond, M., Fournier, D., Malcolm, C. A., Toutant, J.-P. & Arpagaus, M. 1996*b* Existence of two acetylcholinesterases in the mosquito *Culex pipiens* (Diptera: Culicidae). *J. Neurochem.* **67**, 2115–2123.

Cannon, S. C., Brown, R. H. & Corey, D. P. 1991 A sodium channel defect in hyperkalemic periodic paralysis: potassium induced failure of inactivation. *Neuron* **6**, 619–626.

Dong, K. 1997 A single amino acid change in the para sodium channel protein is associated with knockdown-resistance (*kdr*) to pyrethroid insecticides in German cockroaches. *Insect Biochem. Molec. Biol.* **27**, 93–100.

Dong, K. & Scott, J. G. 1994 Linkage of *kdr*-type resistance and the *para*-homologous sodium channel gene in German cockroaches. *Insect Biochem. Molec. Biol.* **24**, 647–654.

Feyereisen, R. 1995 Molecular biology of insecticide resistance. *Toxicol. Lett.* **82/83**, 83–90.

ffrench-Constant, R. H. & Bonning, B. C. 1989 Rapid microtitre plate test distinguishes insecticide resistant acetylcholinesterase genotypes in the mosquitoes *Anopheles albimanus*, *An. nigerrimus* and *Culex pipiens*. *Med. Vet. Entomol.* **3**, 9–16.

ffrench-Constant, R. H., Mortlock, D. P., Shaffer, C. D., MacIntyre, R. J. & Roush, R. T. 1991 Molecular cloning and transformation of cyclodiene resistance in *Drosophila*: an invertebrate GABA$_A$ receptor locus. *Proc. Natn. Acad. Sci. USA* **88**, 7209–7213.

ffrench-Constant, R. H., Roush, R. T. & Carino, F. 1992 *Drosophila* as a tool for investigating the molecular genetics of insecticide resistance. In *Molecular approaches to pure and applied entomology* (ed. M. J. Whitten & J. G. Oakeshott), p. 137. Berlin: Springer.

ffrench-Constant, R. H., Rocheleau, T. A., Steichen, J. C. & Chalmers, A. E. 1993*a* A point mutation in a *Drosophila* GABA receptor confers insecticide resistance. *Nature* **363**, 449–451.

ffrench-Constant, R. H., Steichen, J., Rocheleau, T. A., Aronstein, K. & Roush, R. T. 1993*b* A single amino-acid substitution in a γ-aminobutyric acid subtype A receptor locus associated with cyclodiene insecticide resistance in *Drosophila* populations. *Proc. Natn. Acad. Sci. USA* **90**, 1957–1961.

ffrench-Constant, R. H., Steichen, J. C. & Ode, P. 1993*c* Cyclodiene insecticide resistance in *Drosophila melanogaster* (Meigen) is associated with a temperature sensitive phenotype. *Pestic. Biochem. Physiol.* **46**, 73–77.

ffrench-Constant, R. H., Steichen, J. C. & Brun, L. O. 1994 A molecular diagnostic for endosulfan insecticide resistance in the coffee berry borer *Hypothenemus hampei* (Coleoptera: Scolytidae). *Bull. Entomol. Res.* **84**, 11–16.

ffrench-Constant, R. H., Zhang, H.-G. & Jackson, M. B. 1995 Biophysical analysis of a single amino acid replacement in the resistance to dieldrin γ-aminobutyric acid receptor: novel dual mechanism for cyclodiene insecticide resistance. In *Molecular action of insecticides on ion channels* (ed. J. M. Clarke), pp. 192–204. San Diego, CA: American Chemical Society.

Field, L. M., Anderson, A. P., Denholm, I., Foster, S. P., Harling, Z. K., Javed, N., Martinez-Torres, D., Moores, G. D., Williamson, M. S. & Devonshire, A. L. 1997 Use of biochemical and DNA diagnostics for characterising multiple mechanisms of insecticide resistance in the peach–potato aphid, *Myzus persicae* (Sulzer). *Pestic. Sci.* **51**, 283–289.

Fournier, D., Berrada, S. & Bongibault, V. 1996 Molecular genetics of acetylcholinesterase in insecticide-resistant *Drosophila melanogaster*. In *Molecular genetics and evolution of pesticide resistance* (ed. T. M. Brown), pp. 17–27. Washington: American Chemical Society.

Guerrero, F. D., Jamroz, R. C., Kammlah, D. & Kunz, S. E. 1997 Toxicological and molecular characterization of pyrethroid-resistant horn flies, *Haematobia irritans*: identification of *kdr* and *super-kdr* point mutations. *Insect Biochem. Molec. Biol.* **27**, 745–755.

Haas, R., Marshall, T. L. & Rosenberry, T. L. 1988 *Drosophila* acetylcholinesterase: demonstration of a glycoinositol phospholipid anchor and an endogenous proteolytic cleavage. *Biochemistry* **27**, 6453–6457.

Hall, L. M. & Malcolm, C. A. 1991 The acetylcholinesterase gene of *Anopheles stephensi*. *Cell. Molec. Neurobiol.* **11**, 131–141.

Hall, L. M. & Spierer, P. 1986 The *Ace* locus of *Drosophila melanogaster*: structural gene for acetylcholinesterase with an unusual 5′ leader. *EMBO J.* **5**, 2949–2954.

Hemingway, J. & Georghiou, G. P. 1983 Studies on the acetylcholinesterase of *Anopheles albimanus* resistant and susceptible to organophosphorus and carbamate insecticides. *Pestic. Biochem. Physiol.* **19**, 167–171.

Hemingway, J., Malcolm, C., Kissoon, K., Boddington, R., Curtis, C. & Hill, N. 1985 The biochemistry of insecticide resistance in *Anopheles sacharovi*: comparative studies with a range of insecticide susceptible and resistant *Anopheles* and *Culex* species. *Pestic. Biochem. Physiol.* **24**, 68–76.

Hosie, A. M., Aronstein, K., Sattelle, D. B. & ffrench-Constant, R. 1997 Molecular biology of insect neuronal GABA receptors. *Trends Neurosci.* **20**, 578–583.

Isacoff, E. Y., Jan, Y. N. & Jan, L. Y. 1991 Putative receptor for the cytoplasmic inactivation gate in the *Shaker* K$^+$ channel. *Nature* **353**, 86–90.

Knipple, D. C., Doyle, K. E., Marsella-Herrick, P. A. & Soderlund, D. M. 1994 Tight genetic linkage between the kdr insecticide resistance trait and a voltage-sensitive sodium channel gene in the house fly. *Proc. Natn. Acad. Sci. USA* **91**, 2483–2487.

Kohrman, D. C., Smith, M. R., Goldin, A. L., Harris, J. & Meisler, M. H. 1996 A missense mutation in the sodium channel Scn8a is responsible for cerebellar ataxia in the mouse mutant jolting. *J. Neurosci.* **16**, 5993–5999.

Lee, H.-J., Rocheleau, T., Zhang, H.-G., Jackson, M. B. & ffrench-Constant, R. H. 1993 Expression of a *Drosophila* GABA receptor in a baculovirus insect cell system: functional expression of insecticide susceptible and resistant GABA receptors from the cyclodiene resistance gene *Rdl*. *FEBS Lett.* **335**, 315–318.

Leonard, R. J., Labarca, C. G., Charnet, P., Davidson, N. & Lester, H. A. 1988 Evidence that the M2 membrane-spanning region lines the ion channel pore of the nicotinic receptor. *Science* **242**, 1578–1581.

Loughney, K., Kreber, R. & Ganetzky, B. 1989 Molecular analysis of the *para* locus, a sodium channel gene in *Drosophila*. *Cell* **58**, 1143–1154.

McClatchey, A. I., McKenna-Yasek, D., Cros, D., Worthen, H. G., Kuncl, R. W., DeSilva, S. M., Cornblath, D. R., Gusella, J. F. & Brown, R. H. 1992 Novel mutations in families with unusual and variable disorders of the skeletal muscle sodium channel. *Nat. Genet.* **2**, 148–152.

Martinez-Torres, D., Chandre, F., Williamson, M. S., Darriet, F., Berge, J. B., Devonshire, A. L., Guillet, P., Pasteur, N. & Pauron, D. 1998 Molecular characterization of pyrethroid resistance in the major knockdown resistance (*kdr*) malaria vector *Anopheles gambiae* ss. *Insect Molec. Biol.* **7**, 179–184.

Miyazaki, M., Ohyama, K., Dunlap, D. Y. & Matsumura, F. 1996 Cloning and sequencing of the *para*-type sodium channel gene from susceptible and *kdr*-resistant German cockroaches (*Blattella germanica*) and the housefly (*Musca domestica*). *Molec. Gen. Genet.* **252**, 61–68.

Mutero, A., Pralavorio, M., Bride, J.-M. & Fournier, D. 1994 Resistance-associated point mutations in insecticide-insensitive acetylcholinesterase. *Proc. Natn. Acad. Sci. USA* **91**, 5922–5926.

Newland, C. F. & Cull-Candy, S. G. 1992 On the mechanism of action of picrotoxin on GABA receptor channels in dissociated sympathetic neurones of the rat. *J. Physiol.* **447**, 191–213.

Park, Y. & Taylor, M. F. J. 1997 A novel mutation L1029H in sodium channel gene *hscp* associated with pyrethroid resistance for *Heliothis virescens* (Lepidoptera: Noctuidae). *Insect Biochem. Molec. Biol.* **27**, 9–13.

Pittendrigh, B., Reenan, R., ffrench-Constant, R. H. & Ganetzky, B. 1997 Point mutations in the *Drosophila* sodium channel gene *para* associated with resistance to DDT and pyrethroid insecticides. *Molec. Gen. Genet.* **256**, 602–610.

Ptacek, L. J., George, A. L., Griggs, R. C., Tawil, R., Kallen, R. G., Barchi, R. L., Robertson, M. & Leppert, M. F. 1991 Identification of a mutation in the gene causing hyperkalemic periodic paralysis. *Cell* **67**, 1021–1027.

Raymond, M., Fournier, D., Bride, J.-M., Cuany, A., Berge, J., Magnin, M. & Pasteur, N. 1986 Identification of resistance mechanisms in *Culex pipiens* (Diptera: Culicidae) from southern France: insensitive acetylcholinesterase and detoxifying oxidases. *J. Econ. Entomol.* **79**, 1452–1458.

Rojas, C. V., Wang, J. Z., Schwartz, L. S., Hoffman, E. P., Powell, B. R. & Brown, R. H. 1991 A Met-to-Val mutation in the skeletal muscle Na$^+$ channel alpha-subunit in hyperkalemic periodic paralysis. *Nature* **354**, 387–389.

Severson, D. W., Anthony, N. M., Andreen, O. & ffrench-Constant, R. H. 1997 Molecular mapping of insecticide resistance genes in the yellow fever mosquito (*Aedes aegypti*). *J. Hered.* **88**, 520–524.

Smith, T. J., Lee, S. H., Ingles, P. J., Knipple, D. C. & Soderlund, D. M. 1997 The L1014F point mutation in the housefly *Vssc1* sodium channel confers knockdown resistance to pyrethroids. *Insect Biochem. Molec. Biol.* **27**, 807–812.

Soderlund, D. M. & Bloomquist, J. R. 1989 Neurotoxic actions of pyrethroid insecticides. *A. Rev. Entomol.* **34**, 77–96.

Sussman, J. L., Harel, M., Frolow, F., Oefner, C., Goldman, A., Toker, L. & Silman, I. 1991 Atomic structure of acetylcholinesterase from *Torpedo californica*: a prototypic acetylcholine-binding protein. *Science* **253**, 872–879.

Tang, Z. H. & Cammak, S. L. 1990 Acetylcholinesterase activity in organophosphorus and carbamate resistant and susceptible strains of the *Culex pipiens* complex. *Pestic. Biochem. Physiol.* **37**, 192–199.

Thompson, M., Steichen, J. C. & ffrench-Constant, R. H. 1993 Conservation of cyclodiene insecticide resistance associated mutations in insects. *Insect Molec. Biol.* **2**, 149–154.

Vais, H., Williamson, M. S., Hick, C. A., Eldursi, N., Devonshire, A. L. & Usherwood, P. N. R. 1997 Functional analysis of a rat sodium channel carrying a mutation for insect *knock-down resistance* (*kdr*) to pyrethroids. *FEBS Lett.* **413**, 327–332.

Vaughan, A., Rocheleau, T. & ffrench-Constant, R. 1997 Site-directed mutagenesis of an acetylcholinesterase gene from the yellow fever mosquito *Aedes aegypti* confers insecticide insensitivity. *Expl Parasitol.* **87**, 237–244.

Vaughan, A., Chadee, D. D. & ffrench-Constant, R. 1998 Biochemical monitoring of organophosphorous and carbamate insecticide resistance in *Aedes aegypti* mosquitoes from Trinidad. *Med. Vet. Entomol.* **12**, 318–321.

Wang, Q., Shen, J., Li, Z., Timothy, K., Vincent, G. M., Priori, S. G., Schwartz, P. J. & Keating, M. T. 1995 Cardiac sodium channel mutations in patients with long QT syndrome, an inherited cardiac arrhythmia. *Hum. Molec. Genet.* **4**, 1603–1607.

Williamson, M. S., Denholm, I., Bell, C. A. & Devonshire, A. L. 1993 Knockdown resistance (*kdr*) to DDT and pyrethroid insecticides maps to a sodium channel gene locus in the housefly (*Musca domestica*). *Molec. Gen. Genet.* **240**, 17–22.

Williamson, M. S., Martinez-Torres, D., Hick, C. A. & Devonshire, A. L. 1996 Identification of mutations in the housefly *para*-type sodium channel gene associated with knockdown resistance (*kdr*) to pyrethroid insecticides. *Molec. Gen. Genet.* **252**, 51–60.

Wilson, T. G. 1988 *Drosophila melanogaster* (Diptera: Drosophilidae): a model insect for insecticide resistance studies. *J. Econ. Entomol.* **81**, 22–27.

Yang, N., Ji, J., Zhou, M., Ptacek, L. J., Barchi, R. L., Horn, R. & George, A. L. 1994 Sodium channel mutations in paramyotonia congenita exhibit similar biophysical phenotypes in vitro. *Proc. Natn. Acad. Sci. USA* **91**, 12 785–12 789.

Zhang, H.-G., ffrench-Constant, R. H. & Jackson, M. B. 1994 A unique amino acid of the *Drosophila* GABA receptor with influence on drug sensitivity by two mechanisms. *J. Physiol.* **479**, 65–75.

Zhang, H.-G., Lee, H.-J., Rocheleau, T., ffrench-Constant, R. H. & Jackson, M. B. 1995 Subunit composition determines picrotoxin and bicuculline sensitivity of *Drosophila* GABA receptors. *Molec. Pharmacol.* **48**, 835–840.

Zhu, K. Y. & Clark, J. M. 1997 Validation of a point mutation of acetylcholinesterase in Colorado potato beetle by polymerase chain reaction coupled to enzyme inhibition assay. *Pestic. Biochem. Physiol.* **57**, 28–35.

The role of gene splicing, gene amplification and regulation in mosquito insecticide resistance

Janet Hemingway[1], **Nicola Hawkes**[1], **La-aied Prapanthadara**[2], **K. G. Indrananda Jayawardenal**[1] and **Hilary Ranson**[3]

[1]*School of Pure and Applied Biology, University of Wales Cardiff, PO Box 913, Cardiff CF1 3TL, UK*
[2]*Research Institute of Health Sciences, Chiang Mai University, PO Box 80 CMU, Chiang Mai 50202, Thailand*
[3]*Department of Biological Sciences, University of Notre Dame, Notre Dame, IN 46556, USA*

The primary routes of insecticide resistance in all insects are alterations in the insecticide target sites or changes in the rate at which the insecticide is detoxified. Three enzyme systems, glutathione S-transferases, esterases and monooxygenases, are involved in the detoxification of the four major insecticide classes. These enzymes act by rapidly metabolizing the insecticide to non-toxic products, or by rapidly binding and very slowly turning over the insecticide (sequestration). In *Culex* mosquitoes, the most common organophosphate insecticide resistance mechanism is caused by co-amplification of two esterases. The amplified esterases are differentially regulated, with three times more Est$\beta2^1$ being produced than Est$\alpha2^1$. *Cis*-acting regulatory sequences associated with these esterases are under investigation. All the amplified esterases in different *Culex* species act through sequestration. The rates at which they bind with insecticides are more rapid than those for their non-amplified counterparts in the insecticide-susceptible insects. In contrast, esterase-based organophosphate resistance in *Anopheles* is invariably based on changes in substrate specificities and increased turnover rates of a small subset of insecticides. The up-regulation of both glutathione S-transferases and monooxygenases in resistant mosquitoes is due to the effects of a single major gene in each case. The products of these major genes up-regulate a broad range of enzymes. The diversity of glutathione S-transferases produced by *Anopheles* mosquitoes is increased by the splicing of different 5' ends of genes, with a single 3' end, within one class of this enzyme family. The *trans*-acting regulatory factors responsible for the up-regulation of both the monooxygenase and glutathione S-transferases still need to be identified, but the recent development of molecular tools for positional cloning in *Anopheles gambiae* now makes this possible.

Keywords: mosquitoes; insecticide; gene amplification; gene splicing; *Anopheles*; *Culex*

1. INTRODUCTION

Major mechanisms of insecticide resistance in insects involve either mutation within the target site of the insecticide, or an alteration in the rate of insecticide detoxification. The enzymes involved in this detoxification may be quantitatively and/or qualitatively altered. There are three enzyme groups, esterases, glutathione S-transferases and monooxygenases, involved in metabolic resistance to the four major groups of insecticides. Esterase-based resistance has been reported from more than 30 different medical, veterinary or agricultural insect pests (Hemingway & Karunaratne 1998). In mosquitoes it is the primary mechanism for organophosphorus (OP) insecticide resistance (Bisset *et al.* 1991; Herath *et al.* 1987; Karunaratne *et al.* 1993), and in some cases a secondary mechanism for carbamate resistance (Peiris & Hemingway 1993). Esterases produce a broad spectrum of resistance in many *Culex* species, but in *Anopheles* esterase-based resistance is usually specific to the OP malathion (Hemingway 1983, 1985; Herath *et al.* 1987).

Glutathione S-transferases in mosquitoes commonly confer resistance to the organochlorine insecticide DDT (Prapanthadara *et al.* 1993, 1996), and can act as a secondary OP resistance mechanism (Hemingway *et al.* 1991). In house flies, their role in OP resistance is more widely documented (Clark *et al.* 1984, 1986). DDT resistance in mosquitoes has generally been attributed to a single major-gene effect (Davidson 1963, 1956), although multigenic effects have been suggested in some instances (Lines & Nassor 1991). Reports of monooxygenase-based resistance are relatively rare in mosquitoes, and many of these are based on synergistic effects with piperonyl butoxide, which is not absolutely diagnostic. Pyrethroid resistance in *Anopheles gambiae* in East and West Africa appears to be linked to increased monooxygenase titres, in the latter case combined with an altered target-site mechanism (Brogdon *et al.* 1997). OP resistance in *A. subpictus*, a vector of malaria in Sri Lanka, is also linked to increased monooxygenase titres and higher insecticide metabolic rates (Hemingway *et al.* 1991; Martinez-Torres *et al.* 1998).

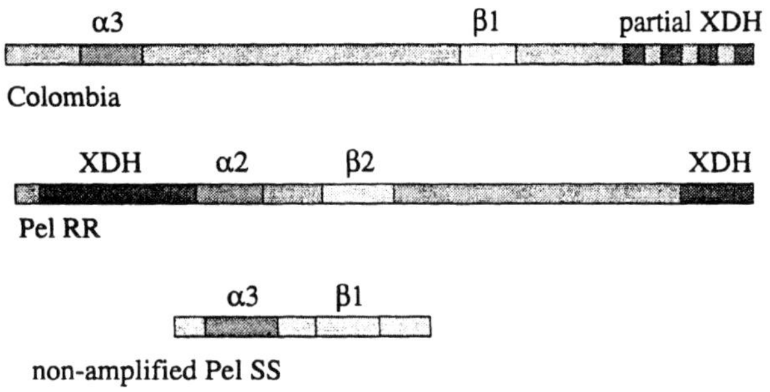

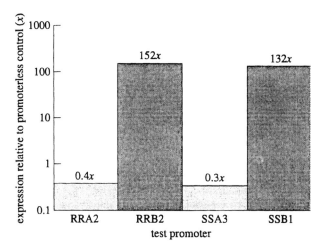

Figure 1. Structure of the amplicons associated with insecticide resistance in *Culex quinquefasciatus* strains from Colombia and Sri Lanka (PelRR) compared with the non-amplified esterase gene arrangement in the insecticide-susceptible strain PelSS. The Est$\alpha 2^l$/$\beta 2^l$ amplicon also has a complete gene with high homology to xanthine dehydrogenase (XDH) (Coleman & Hemingway 1997).

2. THE MOLECULAR BASIS OF METABOLIC RESISTANCE IN MOSQUITOES

(a) *Gene amplification*

The development of resistance to xenobiotics by amplification of the genes involved in their detoxication is common in several organisms. Gene amplification in the insecticide-resistant TEMR strain of the mosquito *Culex quinquefasciatus* was first shown for the Est$\beta 1^l$ esterase. It was originally estimated that there were up to 250 copies of this esterase gene per cell in resistant insects (Mouches *et al.* 1990), but this estimate has recently been revised to approximately 20 copies, i.e. similar to the estimates for *Myzus* esterases (Guillemaud *et al.* 1997). The Est$\beta 1^l$ genes are clustered between the centromere and the apex of chromosome II (Nance *et al.* 1990), and are inherited in a pseudo-monofactorial manner (Peiris & Hemingway 1993). The most common amplified esterases in *Culex* are Est$\alpha 2^l$ and Est$\beta 2^l$, which occur in *ca.* 90% of all the OP-insecticide-resistant *C. quinquefasciatus* strains analysed (Hemingway & Karunaratne 1998). The TEMR Est$\beta 1^l$ and common Est$\beta 2^l$ from numerous strains have 97% identity at the amino-acid level. The high identity suggests that the Estβs are an allelic series from a single locus. The Est$\alpha 2^l$ has approximately 47% deduced amino-acid homology with all the Estβs. This level of homology, along with conserved intron–exon boundaries and the close proximity (1.7 kb) of the two genes in a head-to-head arrangement in the susceptible insects (Vaughan *et al.* 1997), suggests that the two genes arose through an ancient duplication. In resistant insects Est$\alpha 2^l$ and Est$\beta 2^l$ are also in a head-to-head arrangement, but they are *ca.* 2.7 kb apart (Vaughan *et al.* 1997). The increase in the intergenic DNA between the two genes is accounted for by two large (*ca.* 500 bp) and one small insertion in the resistant insects. These insertions introduce DNA motifs that have high homologies to BARBIE boxes, ARE elements and Zeste elements (Vaughan *et al.* 1997; N. Hawkes, unpublished data). The structure of the Est$\alpha 2^l$/EST$\beta 2^l$ amplicon and the related Est$\beta 1$ amplicons compared with the esterase gene arrangements in susceptible insects are given in figure 1.

It has been suggested, on the basis of the identical nature of the amplified Est$\alpha 2^l$ and Est$\beta 2^l$ restriction

Figure 2. Expression rates in a luciferase reporter assay relative to a promoterless control when the intergenic spacer region from resistant (RR) and susceptible (SS) insects is cloned in the orientation of either the Estβ gene or the Estα gene (labelled A and B, respectively).

digest patterns from resistant *Culex* populations worldwide, that amplification is a rare or unique event that occurs primarily through migration (Raymond *et al.* 1991). We now know that amplification of these genes has appeared independently at least five times (Hemingway & Karunaratne 1998), and that resistance is occurring through gene amplification and rapidly spreading by migration. Further evidence that the chromosomal region containing these esterases represents an amplification 'hot-spot' comes from other *Culex* species. In *C. tritaeniorhynchus* the homologous Estβ gene CtrEst$\beta 1^l$ has been amplified and there is no possibility of this having occurred through gene flow between the species (Karunaratne *et al.* 1998).

In *Anopheles stephensi* there are at least three enzymes that are able to metabolize the OP malathion in resistant insects. None of these esterase genes are amplified and they are all present in very low quantities in the resistant and susceptible insects, conferring resistance through efficient insecticide metabolism (K. G. I. Jayawardena and J. Hemingway, unpublished data).

(b) *Gene expression*

Increased gene expression, rather than gene amplification, is the primary molecular basis of glutathione S-transferase and monooxygenase-based resistance in mosquitoes. However, gene amplification and elevated expression are not mutually exclusive. The Est$\alpha 2^l$ and Est$\alpha 2^l$ genes from *Culex* appear to be both amplified and increased in expression. The two genes are present in a 1:1 stoichiometry, being co-amplified, but approximately three times more Est$\beta 2^l$ than Est$\alpha 2^l$ is obtained from protein purifications of resistant insect homogenates (Karunaratne 1994). This difference may reflect differential protein or mRNA stability, or result from variations in the efficiencies of the two promoters, which are both contained within the intergenic spacer (figure 1). We are currently characterizing the Est$\alpha 2^l$ and Est$\beta 2^l$ promoters, and have cloned the intergenic spacer in both orientations upstream of the reporter gene luciferase. The

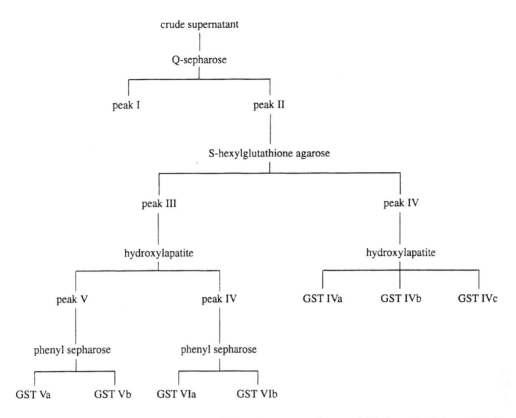

Figure 3. Schematic representation of the purification of glutathione S-transferases (GST) from *Anopheles gambiae*. The final peaks of GST activity still all contained multiple protein bands.

Table 1. *DDT dehydrochlorinase activity*[a] *exhibited by various peaks of glutathione S-tranferase activity partially purified from* A. gambiae, *as described in figure 3*

(S, susceptible strain enzymes; R, resistant strain enzymes.)

GSTs	nmole DDE mg^{-1}		nmole/unit GST^{b} DDE activity		nmole DDE g^{-1} larvae	
	S	R	S	R	S	R
IVa	—	278.0	<0.3	6.1	—	14.6
IVb	—	912.5	<0.3	22.6	—	15.6
IVc	—	22.0	<0.3	2.2	—	1.5
Va	173.3	564.7	144.5	241.3	48.4	344.9
Vb	65.6	765.6	121.5	243.1	7.6	68.1
VIa	27.4	50.5	94.6	112.2	4.2	39.6
VIb	9.8	178.6	98.3	235.1	0.9	52.4

[a] DDT-dehydrochlorinase activity is defined as nmole DDE formation per two hours.
[b] A unit of GST activity is defined as μmole min^{-1} mg^{-1} with CDNB as the substrate.

resultant constructs have been transfected into a range of insect and mammalian cell lines. Inserting the spacer at the same site, but in different orientations, reproducibly generates luciferase expression from the Estβ2[l] promoter many times greater than from the Estα2[l] orientation. This is true from both the resistant (2.7 kb) and susceptible (1.7 kb) spacers (figure 2). The differences in promoter strength may reflect differences in the relative locations of significant regulatory elements. Further studies are in progress.

In the malaria vector, *Anopheles gambiae*, resistance to DDT in both larvae and adults is conferred by increased levels of many glutathione S-transferases (GSTs). Resistance in the two life stages is conferred by different genes, although the end result of both is a measurable increase in GST activity and DDT dehydrochlorination. Resistance has been studied in greatest detail in larvae. Resistant insects have increased levels of DDT dehydrochlorinase activity associated with seven partially purified peaks of GST activity (figure 3; and table 1) (Prapanthadara *et al.* 1993, 1995). There is an enormous diversity of GSTs found in both resistant and susceptible insects.

At present two broad classes of GSTs have been cloned from insects. Representatives from both classes have been cloned from *A. gambiae* (although the single insect class II GST cloned from *A. gambiae* is not expressed in larvae). All three class I GSTs cloned are able to use DDT as a substrate (Ranson *et al.* 1997*a,b*). Antisera raised to these expressed GSTs indicates that all of them belong to the peak IV GSTs represented in figure 3. We do not have, as yet, have any molecular data on the GSTs from peaks V and VI, but our current information suggests that they belong to GST classes that have not so far been characterized from any insect. The simplest hypothesis for the molecular basis of this GST-based resistance and for the similar organization of the monooxygenase-based pyrethroid resistance is that *trans*-acting regulators are involved in the up-regulation of these enzyme families. We are currently employing a positional cloning approach, in collaboration with Professor F. Collins, USA, to identify these regulator genes. The positional cloning takes advantage of the high-density microsatellite marker genetic map, *in situ* hybridization, polytene chromosome microdissection, and a BAC library, which have all been recently developed for *A. gambiae*.

(c) *Gene splicing*

Initial biochemical work on GSTs from *A. gambiae* demonstrated the diversity of GST enzymes present in this insect. The molecular work undertaken subsequent to this showed that the class I GSTs were all at a single chromosome location. Extensive sequencing of stretches of DNA at this location revealed a full-length intron-less gene, as occurs in *Drosophila*, and numerous apparent 5′ truncated pseudogenes. Reverse transcriptase-polymerase chain reaction and Southern blot analysis has demonstrated that these are not pseudogenes, but are actively transcribed with the splicing of different 5′ exons to a single 3′ exon, occurring to produce a diversity of GSTs in both resistant and susceptible *A. gambiae* (Ranson *et al.* 1998).

The next decade should see extensive progress in our understanding of metabolically based insecticide resistance in insect pests, allowing for the development of new control methods to allow us to counteract this rapidly changing evolutionary phenomenon.

The unpublished work cited in this paper was carried out with project grant funding to J.H. from the Medical Research Council and a Wellcome Prize studentship to H.R.

REFERENCES

Bisset, J. A., Rodriguez, M. M., Hemingway, J., Diaz, C., Small, G. J. & Ortiz, E. 1991 Malathion and pyrethroid resistance in *Culex quinquefasciatus* from Cuba: efficacy of pirimiphos-methyl in the presence of at least three resistance mechanisms. *Med. Vet. Entomol.* **5**, 223–228.

Brogdon, W. G., McAllister, J. C. & Vulule, J. 1997 Heme peroxidase activity measured in single mosquitoes identifies individuals expressing the elevated oxidase mechanism for insecticide resistance. *J. Am. Mosq. Cont. Ass.* **13**, 233–237.

Clark, A. G., Shamaan, N. A., Dauterman, W. C. & Hayaoka, T. 1984 Characterization of multiple glutathione transferases from the housefly, *Musca domestica* (L). *Pestic. Biochem. Physiol.* **22**, 51–59.

Clark, A. G., Shamaan, N. A., Sinclair, M. D. & Dauterman, W. C. 1986 Insecticide metabolism by multiple glutathione S-transferases in two strains of the house fly, *Musca domestica* (L). *Pestic. Biochem. Physiol.* **25**, 169–175.

Coleman, M. & Hemingway, J. 1997 Amplification of a xanthine dehydrogenase gene is associated with insecticide resistance in the common house mosquito *Culex quinquefasciatus*. *Biochem. Soc. Trans.* **25**, 526.

Davidson, G. 1956 Insecticide resistance in *Anopheles gambiae* Giles a case of simple Mendelian inheritance. *Nature* **178**, 863.

Davidson, G. 1963 DDT-resistance and dieldrin-resistance in *Anopheles albimanus*. *Bull. Wld Hlth Org.* **28**, 25–33.

Guillemaud, T., Makate, N., Raymond, M., Hirst, B. & Callaghan, A. 1997 Esterase gene amplification in *Culex pipiens*. *Insect. Molec. Biol.* **6**, 319–327.

Hemingway, J. 1983 The genetics of malathion resistance in *Anopheles stephensi* from Pakistan. *Trans. R. Soc. Trop. Med. Hyg.* **77**, 106–108.

Hemingway, J. 1985 Malathion carboxylesterase enzymes in *Anopheles arabiensis* from Sudan. *Pestic. Biochem. Physiol.* **23**, 309–313.

Hemingway, J. & Karunaratne, S. H. P. P. 1998 Mosquito carboxylesterases: a review of the molecular biology and biochemistry of a major insecticide resistance mechanism. *Med. Vet. Entomol.* **12**, 1–12.

Hemingway, J., Miyamoto, J. & Herath, P. R. J. 1991 A possible novel link between organophosphorus and DDT insecticide resistance genes in *Anopheles*: supporting evidence from fenitrothion metabolism studies. *Pestic. Biochem. Physiol.* **39**, 49–56.

Herath, P. R. J., Hemingway, J., Weerasinghe, I. S. & Jayawardena, K. G. I. 1987 The detection and characterization of malathion resistance in field populations of *Anopheles culicifacies* B in Sri Lanka. *Pestic. Biochem. Physiol.* **29**, 157–162.

Karunaratne, S. H. P. P. 1994 Characterization of multiple variants of carboxylesterases which are involved in insecticide resistance in the mosquito *Culex quinquefasciatus*. PhD thesis, University of London.

Karunaratne, S. H. P. P., Jayawardena, K. G. I., Hemingway, J. & Ketterman, A. J. 1993 Characterization of a B-type esterase involved in insecticide resistance from the mosquito *Culex quinquefasciatus*. *Biochem. J.* **294**, 575–579.

Karunaratne, S. H. P. P., Vaughan, A., Paton, M. G. & Hemingway, J. 1998 Amplification of a serine esterase gene is involved in insecticide resistance in Sri Lankan *Culex tritaeniorhynchus*. *Insect. Molec. Biol.* **7**, 307–315.

Lines, J. D. & Nassor, N. S. 1991 DDT resistance in *Anopheles gambiae* declines with mosquito age. *Med. Vet. Entomol.* **5**, 261–265.

Martinez-Torres, D., Chandre, F., Williamson, M. S., Darriet, F., Berge, J. B., Devonshire, A. L., Guillet, P., Pasteur, N. & Pauron, D. 1998 Molecular characterisation of pyrethroid resistance in the malaria vector *Anopheles gambiae* s.s. *Insect. Molec. Biol.* **7**, 179–184.

Mouches, C. (and 10 others) 1990 Characterization of amplification core and esterase B1 gene responsible for insecticide resistance in *Culex*. *Proc. Natn. Acad. Sci. USA* **87**, 2574–2578.

Nance, E., Heyse, D., Britton-Davidian, J. & Pasteur, N. 1990 Chromosomal organisation of the amplified esterase B1 gene in organophosphate-resistant *Culex pipiens qinquefasciatus* Say (Diptera: Culicidae). *Genome* **33**, 148–152.

Peiris, H. T. R. & Hemingway, J. 1993 Characterization and inheritance of elevated esterases in organophosphorus and carbamate insecticide resistant *Culex quinquefasciatus* (Diptera:Culicidae) from Sri Lanka. *Bull. Entomol. Res.* **83**, 127–132.

Prapanthadara, L., Hemingway, J. & Ketterman, A. J. 1993 Partial purification and characterization of glutathione S-transferase involved in DDT resistance from the mosquito *Anopheles gambiae*. *Pestic. Biochem. Physiol.* **47**, 119–133.

Prapanthadara, L., Hemingway, J. & Ketterman, A. J. 1995 DDT-resistance in *Anopheles gambiae* Giles from Zanzibar Tanzania, based on increased DDT-dehydrochlorinase activity of glutathione S-transferases. *Bull. Entomol. Res.* **85**, 267–274.

Prapanthadara, L., Koottathep, S., Promtet, N., Hemingway, J. & Ketterman, A. J. 1996 Purification and characterization of a major glutathione S-transferase from the mosquito *Anopheles dirus* (species B). *Insect. Biochem. Molec. Biol.* **26**, 277–285.

Ranson, H., Cornel, A. J., Fournier, D., Vaughan, A. & Hemingway, J. 1997*a* Cloning and localisation of a glutathione S-transferase class I gene from *Anopheles gambiae*. *J. Biol. Chem.* **272**, 5464–5468.

Ranson, H., Prapanthadara, L. & Hemingway, J. 1997*b* Cloning and characterisation of two glutathione S-transferases from a DDT resistant strain of *Anopheles gambiae. Biochem. J.* **324**, 97–102.

Ranson, H., Collins, F. H. & Hemingway, J. 1998 The role of mRNA splicing in generating heterogeneity within the *Anopheles gambiae* class I glutathione S-transferase gene family. *Proc. Natn. Acad. Sci. USA.* (In the press.)

Raymond, M., Callaghan, A., Fort, P. & Pasteur, N. 1991 Worldwide migration of amplified insecticide resistance genes in mosquitoes. *Nature* **350**, 151–153.

Vaughan, A., Hawkes, N. & Hemingway, J. 1997 Co-amplification explains linkage disequilibrium of two mosquito esterase genes in insecticide resistant *Culex quinquefasciatus. Biochem. J.* **325**, 359–365.

Cytochrome P450 monooxygenases and insecticide resistance in insects

Jean-Baptiste Bergé[1], René Feyereisen[2] and Marcel Amichot[1]

[1]*INRA, 123 Boulevard Francis Meilland, BP2078, 06606 Antibes Cedex, France*
[2]*Department of Entomology, Forbes 410, PO Box 210036, University of Arizona, Tucson, AZ 85721-0036, USA*

Cytochrome P450 monooxygenases are involved in many cases of resistance of insects to insecticides. Resistance has long been associated with an increase in monooxygenase activities and with an increase in cytochrome P450 content. However, this increase does not always account for all of the resistance. In *Drosophila melanogaster*, we have shown that the overproduction of cytochrome P450 can be lost by the fly without a corresponding complete loss of resistance. These results prompted the sequencing of a cytochrome P450 candidate for resistance in resistant and susceptible flies. Several mutations leading to amino-acid substitutions have been detected in the P450 gene *CYP6A2* of a resistant strain. The location of these mutations in a model of the 3D structure of the CYP6A2 protein suggested that some of them may be important for enzyme activity of this molecule. This has been verified by heterologous expression of wild-type and mutated cDNA in *Escherichia coli*. When other resistance mechanisms are considered, relatively few genetic mutations are involved in insecticide resistance, and this has led to an optimistic view of the management of resistance. Our observations compel us to survey in more detail the genetic diversity of cytochrome P450 genes and alleles involved in resistance.

Keywords: cytochrome P450; insecticide metabolism; *Drosophila melanogaster*; overexpression; point mutations

1. GENERAL INFORMATION ON MONOOXYGENASE ACTIVITIES AND THE P450 CYTOCHROMES

The P450 monooxygenases are ubiquitous enzymes, found from bacteria to mammals. They are involved in endogenous metabolism as well as in the metabolism of xenobiotics. For example, in insects these activities are essential for the synthesis and the degradation of the steroid moulting hormones and juvenile hormones and also in the metabolism of pheromones. The P450 enzymes are also important for the adaptative mechanisms of insects to the toxic chemicals synthesized by their host plants (Gould 1984). This adaptation is notable for the fact that the biosynthesis of these enzymes can be induced by the presence of the toxins in the food (Frank & Fogleman 1992; Berenbaum *et al.* 1990; Hung *et al.* 1995). We also know that P450 monooxygenase activities can be involved in the metabolism of virtually all insecticides, leading to an activation of the molecule or, more generally, to a detoxification (Wilkinson & Brattsten 1972; Hodgson 1985; Agosin 1985). For some insects, this detoxification is so active (Taylor & Feyereisen 1996) that the insecticide does not reach its molecular target before being metabolized and degraded by these enzymes: such individuals are resistant to insecticides.

P450 enzymes bind molecular oxygen and receive electrons from NADPH to introduce an oxygen atom into the substrate and to form water with the other oxygen atom according to the reaction:

$$\text{substrate(S)} + (\text{NADPH} + \text{H}^+) + \text{O}_2 \rightarrow \text{S(O)} + \text{NADP}^+ + \text{H}_2\text{O}.$$

The electrons necessary for this reaction are transferred from nicotinamide-adenine dinucleotide phosphate (NADPH) on the 'substrate–P450' complex by an NADPH cytochrome P450 reductase, but this reaction can also be stimulated by cytochrome b_5 (Guzov *et al.* 1996; Megias *et al.* 1984; Zhang & Scott 1996). The stability of the initial product [S(O)] can vary, leading to final overall reactions as diverse as hydroxylation, epoxidation, O-, N-, and S- dealkylations, N- and S- oxidations and to such various chemical reactions, and products that these enzymes have been called 'diversozymes' (Coon *et al.* 1996). The key protein of this enzymatic system is in each case a cytochrome P450 that is responsible for the specificity of the reaction. This protein has an absorption at 450 nm when reduced and saturated with CO, hence its name (Omura & Sato 1964). Comprehensive reviews on P450 from insects have been published (Wilkinson & Brattsten 1972; Hodgson 1985; Agosin 1985), and an updated review will be published soon (Feyereisen 1999). If P450 monooxygenase activities are exerted on such a significant diversity of substrates (steroids, juvenile hormone, hydrocarbons, pesticides, etc.), it is because there is a high number of cytochromes P450 in each individual. The P450s certainly constitute one of the most important superfamilies of proteins, considering the large number of forms. To cope with such a diversity it was necessary to adopt a nomenclature based on sequences homologies of P450 and hence on phylogeny (Nelson *et al.* 1996). This nomenclature, now universally accepted, designates all gene members of the P450 superfamily with a CYP prefix, followed by a numeral for the family, a letter for the subfamily, and a numeral for the individual gene. All members of a family share more than 40% identity at the amino-acid sequence level, and members of a subfamily share more than 55% identity. Genes are described in italics, whereas the gene product, mRNA, and enzyme are in capitals. To date, insect P450s have been assigned to six CYP families; five are insect-specific (CYP6, 9, 12, 18 and 28), and one, CYP4, is shared with sequences from other organisms. Extrapolations based on the known P450 and on a screen of the currently available single transcribed sequences (STSs) and expressed sequence tags (ESTs) of *Drosophila melanogaster* lead to the idea that the total number of P450s in this species will be between 60 and 100 (Feyereisen 1999).

2. CYTOCHROME P450 AND THE RESISTANCE OF INSECTS TO INSECTICIDES: HIGHLIGHTS

It is well established that many cases of metabolic resistance of insects to insecticides are the result of enhanced P450 activities. The involvement of P450 enzymes in resistance can be shown by several methods. P450 monooxygenase inhibitors such as piperonyl butoxide are most commonly used. Treatment of resistant insects by piperonyl butoxide can result in a complete loss of resistance, indicating that resistance is due only to P450 activity. Such a conclusion assumes that the effects of piperonyl butoxide are restricted to P450 inhibition, an assumption that is not always correct. Confirmation with other P450 synergists chemically unrelated to piperonyl butoxide (e.g. imidazole- or propynylether-type synergists) is usually in order. In the majority of the cases, however, resistance is due to several mechanisms and the treatment with the P450 synergist does not restore complete susceptibility. Thus in LearnPyrR, a pyrethroid-resistant strain of house flies, the resistance factor is 6000, but when the flies are treated with piperonyl butoxide this factor decreases to no more than 32 owing to residual resistance that involves a decrease in the penetration kinetics and modification of the target (Scott & Georghiou 1986). A more direct way to show the intervention of P450 in resistance to insecticides is to compare directly the NADPH-dependent metabolism of the insecticide in resistant and susceptible strains. This direct method is not often used because it requires radiolabelled molecules that are not always available to the researchers. In the case of the DDT-resistant strain RDDT^R of *Drosophila*, the NADPH-dependent metabolism of DDT is ten times higher than that of a susceptible strain (Cuany *et al.* 1990). Whatever the method of characterizing resistance, one can say that P450-dependent resistance has been reported for most insecticide classes and in most arthropod pest species, highlighting the need to obtain a good knowledge of this resistance mechanism.

3. P450 MONOOXYGENASE ACTIVITIES AND CONTENT OF P450 IN RESISTANT STRAINS

The total P450 content can be measured by optical spectroscopy by means of the absorption spectrum of P450 reduced and saturated with CO (Omura & Sato 1964). This content has been compared in several resistant and susceptible strains of the house fly (Scott *et*

al. 1990). For all the strains that have resistance synergized by piperonyl butoxide there is an increase in the total content of cytochrome P450. This phenomenon has been known for a long time (Hodgson 1985; Agosin 1985; DeVries & Georghiou 1981; Vincent *et al.* 1985; Dyte 1972; Cohen 1982). Interestingly, it has been observed that, in addition to the increase in P450, there is also an increase in P450 reductase and cytochrome b_5 in some resistant strains (Scott & Georghiou 1986; Scott *et al.* 1990). This measurement of the total increase in P450 is an underestimate of the increase in specific forms of P450. The P450 overproduced in the resistant house fly strain LearnPyrR was purified and a specific antibody was produced (Wheelock & Scott 1990). An immunoassay of the overproduced P450 from LearnPyrR (called P450lpr) has shown that it accounts for 68% of the total P450, 44 times more than the level of P450lpr in the susceptible strain. Similar results were obtained on the RDDT^R strain of *Drosophila*, which has 40 times more CYP6A2 protein than the susceptible strain (M. Amichot and A. Brun, unpublished results).

The first cloning of a P450 cDNA from insects, CYP6A1 from the house fly, was via an expression library of cDNA obtained from the resistant Rutgers strain overexpressing P450 (Feyereisen *et al.* 1989). By means of several PCR methods many other P450s have been cloned, making probes available to show that the mRNA of several P450s is constitutively overproduced in resistant strains: CYP6A1 is overproduced in the resistant Rutgers strain (Carino *et al.* 1992, 1994) and CYP6D1 in LearnPyrR (Scott *et al.* 1996); CYP6A2 is overproduced in resistant strains of *Drosophila* (Waters *et al.* 1992; Brun *et al.* 1996), whereas CYP6A9 is overproduced in other resistant strains of this species (Maitra *et al.* 1996); CYP6B2 is overproduced in *Helicoverpa* (Xiao-Ping & Hobbs 1995), as is CYP4G8 (Pittendrigh *et al.* 1997) and CYP9A1 in *Heliothis virescens* (Rose *et al.* 1997). In fact, several P450s can be overproduced together in an individual, e.g. CYP6A2 and CYP4E2 in the RDDT^R strain of *Drosophila* (Amichot *et al.* 1994), and CYP6A1 and CYP6D1 in LearnPyrR house flies (Carino *et al.* 1992). This overproduction can be due to an overexpression of the gene encoding these proteins, but a stabilization of the corresponding mRNA or protein cannot be excluded; to date, no gene amplification has been demonstrated in strains overproducing P450. It has been shown by mRNA *in situ* hybridization (Brun *et al.* 1996) that overproduction does not modify the spatial- and tissue-specific expression of CYP6A2, which is specifically expressed in the proximal gut, in the Malpighian tubules and in the subcuticular fat bodies. This overproduction of P450 must be involved in the resistance; indeed, when CYP6A1, CYP12A1 (Feyereisen 1999) and CYP6A2 (Dunkov *et al.* 1997) were expressed in *E. coli* or in baculovirus-infected cells, these P450 could cleave oxidatively the ester bond of diazinon, a reaction that represents a detoxification of the molecule.

The genetic mechanism responsible for P450 constitutive overproduction is not well understood; however, it is known that in the overproducing strains of the house fly and *Drosophila* there is an interference with the process of induction of these proteins by phenobarbital (Carino *et al.* 1994; Brun *et al.* 1996). At present, the best-characterized model is that of the Rutgers house fly strain, resistant to organophosphates. In this strain, resistance is at least associated with chromosome 2 (Plapp 1984), but it is known that the structural gene for CYP6A1 is on chromosome 5 (Cohen *et al.* 1994). Similar results were obtained with CYP6D1, which is located on chromosome 1, whereas its expression is regulated by a factor on chromosome 2 (Liu & Scott 1996*a*), on which a factor regulating sensitivity to phenobarbital has also been reported (Liu & Scott 1997). Thus, at least in the house fly, there would be obviously a *trans* genetic factor relative to CYP6A1 that would control its expression and whose modification would be at the origin of the switch from low constitutive expression in the insecticide-susceptible flies to a constitutive overproduction in the resistant flies. In *Drosophila*, the data are less clear, but genetic data suggest that there is also a *trans* regulation of the overexpression of P450 (Waters & Nix 1988; Houpt *et al.* 1988). Molecular data lead to the same conclusion concerning the overexpression of CYP6A2, CYP4E2 and CYP6A9 (Maitra *et al.* 1996; Amichot *et al.* 1994).

This increase in the content of a component of the P450 system results in an increase in the enzymatic activity for insecticides in resistant insects. However, at least in *Drosophila*, one can also observe an increase in activity on substrates as varied as ethoxycoumarin, ethoxyresorufin, ecdysteroids, testosterone, and lauric acid and some of its unsaturated derivatives (Cuany *et al.* 1990). This diversity of substrates metabolized in resistant insects suggests that several P450s are overproduced in the resistant strains of *Drosophila*. However, we cannot eliminate the possibility that an overproduced P450 in the resistant strain has, in addition, a broader substrate specificity when compared with that of the allele present in the susceptible strain. A practical and rapid system to measure P450 activity via O-de-ethylation of 7-ethoxycoumarin (ECOD) in a single fruit fly was developed (deSousa *et al.* 1995). This technique, applied to determine ECOD activity in individuals from wild populations of *Drosophila* (Bride *et al.* 1997) and *Cydia pomonella* (Sauphanor *et al.* 1998), showed that this activity is well correlated with resistance level and that, compared to strains selected in laboratory, the standard deviation of measurements is much more important in wild populations than in the populations reared for a long time in the laboratory. In these wild populations there are probably various types of individuals, homozygotes and heterozygotes for the overexpression of this activity.

4. RESISTANCE TO INSECTICIDES AND AMINO-ACIDS SUBSTITUTIONS IN CYP6A2

The P450-dependent resistance cannot always be fully accounted for by an increase in the content of cytochrome P450. For example, no increase in CYP6A1 mRNA content has been observed in some strains of flies resistant to insecticides by a mechanism that can be inhibited by piperonyl butoxide (Carino *et al.* 1992). In the strain LearnPyrR, it is also impossible to correlate the piperonyl butoxide-dependent resistance to permethrin (resistance factor greater than 1000) (Sauphanor *et al.* 1998) with the ninefold overproduction of CYP6D1 protein (Scott *et al.* 1996). These observations, and the fact that in the RDDT[R]-resistant strain of *Drosophila* the

resistance is polygenic, led us to attempt a separation of the various resistance factors via backcrosses between the resistant strain RDDT[R] and the susceptible strain 88100. The progeny of each backcross was selected by DDT by tarsal contact at $50 \, \mathrm{nmol \, cm^{-2}}$. After 15 selective backcrosses the strain obtained (called 152) metabolized DDT more intensely than the susceptible strain, and this metabolism is NADPH-dependent. The 152 strain has a monogenic P450-dependent resistance to DDT, with an LC50 of $60 \, \mathrm{nmol \, cm^{-2}}$ for DDT (RDDT[R] has an LC50 higher than $1 \, \mathrm{mmol \, cm^{-2}}$). Moreover, the 152 strain does not overproduce CYP6A2. Using this strain, the resistance factor was mapped to the approximate chromosome location 55–56. Owing to the imprecision of the correspondence between the genetic localization and the mapping determined via *in situ* hybridization, the 55–56 locus could well correspond to the 43A band on which CYP6A2 has been localized on polytene chromosomes. CYP9B1, which could be another candidate for resistance, also maps in this region but it is not overexpressed in the RDDT[R] strain. Moreover, the homology between house fly CYP6A1 and *Drosophila* CYP6A2 is shown by the following facts (Dunkov *et al.* 1997).

1. There is a high degree (49%) of sequence identity for these members of the CYP6A subfamily.
2. They are localized on homologous chromosomes.
3. They have only one intron located at the same place.
4. They are both induced by phenobarbital and their promoter has characteristic BARBIE box sequences.
5. They are overexpressed in resistant strains; this is not the case for other CYP6As in the house fly.
6. Their expression is under the control of a factor found on chromosome II for the house fly and chromosome 3 for *Drosophila*, which are homologous.
7. They both metabolize diazinon and cyclodienes.

This probable orthology reinforces the idea that CYP6A1 and CYP6A2 are both contributing to resistance. The comparison of the sequences between CYP6A2 from a susceptible strain of *Drosophila* and CYP6A2 from RDDT[R] or strain 152 shows that there are three amino-acid substitutions, R335S, L336V and V476L. Modelling of CYP6A2 based on sequence homologies with several crystallized P450s revealed that these three mutations may have an effect on the structure of the active site of CYP6A2. We thus expressed in *E. coli* a wild-type CYP6A2 and this same P450 mutagenized in order to introduce the mutations alone or in combination. The results to date show that these mutations do not modify the activity of CYP6A2 for testosterone, but that there is an increase in activity for 7-ethoxycoumarin, 7-benzoyloxycoumarin and especially DDT, hydroxylated to the non-insecticidal dicofol.

5. QUESTIONS AND WORKING HYPOTHESES AS CONCLUSIONS

It seems that CYP6A1 and CYP6A2 are very significant factors for P450-dependent insecticide resistance in house flies and *Drosophila*, respectively. In the latter case, resistance probably would results from a combination of overproduction and amino-acid substitution, which would lead to an overproduced cytochrome P450 with a good catalytic efficiency with respect to the insecticide. However, many questions still remain outstanding. The overproduction of P450 appears unstable, relatively easily lost in the absence of selection, and one wonders why this is so? Does this reflect the relative importance in resistance of mutations causing overproduction and amino-acid substitution mutations? Some studies suggest that P450 overproduction decreases the fitness of individuals, which is logical as it is known that the overproduced P450 can metabolize hormonal endogenous molecules (Cuany *et al.* 1990). It is possible that amino-acid substitutions may involve less disturbances to the fitness of the individual that carries them. Once fixed in populations, these substitutions, if they confer significant resistance, would facilitate the loss of overproduction, a form of genetic succession (Taylor & Feyereisen 1996). What is the gene regulating the overexpression of P450 in the resistant insects? This is still unknown. Finally, it remains unclear how many different P450s participate in resistance in a given strain, and how many amino-acid substitutions of importance in resistance will be found in P450s?

We now have many tools that should enable us to answer these questions. In any event, it is only then that we will be able to consider seriously the possibilities of monitoring accurately each resistant allele of P450 and of managing their spread in wild populations of agricultural pests or vectors of disease.

REFERENCES

Agosin, M. 1985 Role of microsomal oxidations in insecticide degradation. *Comp. Insect Physiol. Biochem. Physiol.* **12**, 647–712.

Amichot, M., Brun, A., Cuany, A., Helvig, C., Salaun, J. P., Durst, F. & Bergé, J.-B. 1994 Expression study of *CYP* genes in *Drosophila* strains resistant or susceptible to insecticides. In *Cytochrome P450. 8th Int. Conf.* (ed. M. C. Lechner), pp. 689–692. Paris: Eurotext–John Libbey.

Berenbaum, M. R., Cohen, M. B. & Shuler, M. A. 1990 Cytochrome P450 in plant–insect interactions: inductions and deductions. In *Molecular insect science* (ed. H. H. Hagedorn, J. G. Hildebrand, M. G. Kidwell & J. H. Law), pp. 257–262. New York: Plenum.

Bride, J. M., Cuany, A., Amichot, M., Brun, A., Babault, M., Le Mouel, T., De Sousa, G., Rahmani, R. & Berge, J. B. 1997 Cytochrome P450—field insecticide tolerance and development of laboratory resistance in grape-vine populations of *Drosophila melanogaster* (Diptera: Drosophilidae). *J. Econ. Entomol.* **90**, 1514–1520.

Brun, A., Cuany, A., Le Mouel, T., Berge, J. B. & Amichot, M. 1996 Inducibility of the *Drosophila melanogaster* cytochrome P450 gene, *CYP6A2*, by phenobarbital in insecticide susceptible or resistant strains. *Insect Biochem. Molec. Biol.* **26**, 697–703.

Carino, F. A., Koener, J. F., Plapp, F. W. & Feyereisen, R. 1992 Expression of the cytochrome P450 gene *CYP6A1* in the housefly, *Musca domestica*. *ACS Symp. Ser.* **505**, 31–40.

Carino, F. A., Koener, J. F., Plapp, F. W. & Feyereisen, R. 1994 Constitutive overexpression of the cytochrome P450 gene *CYP6A1* in a house fly strain with metabolic resistance to insecticides. *Insect Biochem. Molec. Biol.* **24**, 411–418.

Cohen, E. 1982 Studies on several microsomal enzymes in two strains of *Tribolium castaneum* (Tenebrionidae: Coleoptera). *Comp. Biochem. Physiol.* C **71**, 123–131.

Cohen, M. B., Koener, J. F. & Feyereisen, R. 1994 Structure and chromosomal localization of *CYP6A1*, a cytochrome P450 encoding gene from the house fly. *Gene* **146**, 267–272.

Coon, M. J., Vaz, A. D. & Bestervelt, L. L. 1996 Peroxidative reactions of diversozymes. *FASEB J.* **10**, 428–434.

Cuany, A., Pralavorio, M., Pauron, D., Bergé, J.-B., Fournier, D., Blais, C., Lafont, R., Salaun, J. P., Weissbart, D., Larroque, C. & Lange, R. 1990 Characterization of microsomal oxidative activities in a wild-type and in a DDT resistant strain of *Drosophila melanogaster*. *Pestic. Biochem. Physiol.* **37**, 293–302.

deSousa, G., Cuany, A., Amichot, M., Rahmani, G. & Bergé, J.-B. 1995 A fluorometric method for measuring ECOD activity on individual abdomen of *Drosophila melanogaster*: application to the study on resistance of insects to insecticides. *Analyt. Biochem.* **229**, 86–91.

DeVries, D. H. & Georghiou, G. P. 1981 Absence of enhanced detoxication of permethrin in pyrethroid-resistant house flies. *Pestic. Biochem. Physiol.* **15**, 242–250.

Dunkov, B. C., Guzov, V. M., Mocelin, G., Shotkoski, F., Brun, A., Amichot, M., ffrench-Constant, R. H. & Feyereisen, R. 1997 The *Drosophila* cytochrome P450 gene *Cyp6a2*: structure, localization, heterologous expression and induction by phenobarbital. *DNA Cell Biol.* **16**, 1345–1356.

Dyte, C. E. 1972 Resistance to synthetic juvenile hormone in a strain of flour beetle, *Tribolium castaneum*. *Nature* **238**, 48–51.

Feyereisen, R. 1999 Insect P450 enzymes. *A. Rev. Entomol.* (In the press.)

Feyereisen, R., Koener, J. F., Farnsworth, D. E. & Nebert, D. W. 1989 Isolation and sequence of cDNA encoding a cytochrome P450 from an insecticide-resistant strain of the house fly, *Musca domestica*. *Proc. Natn. Acad. Sci. USA* **86**, 1465–1469.

Frank, M. R. & Fogleman, J. C. 1992 Involvement of cytochrome P450 in host-plant utilization by Sonoran Desert *Drosophila*. *Proc. Natn. Acad. Sci. USA* **89**, 11 998–12 002.

Gould, F. 1984 Mixed function oxidases and herbivore polyphagy: the devil's advocate position. *Ecol. Entomol.* **9**, 29–34.

Guzov, V., Houston, H., Murataliev, M. B., Walker, F. A. & Feyereisen, R. 1996 Molecular cloning, overexpression in *E. coli*, structural and functional characterization of house fly cytochrome b5. *J. Biol. Chem.* **271**, 26 637–26 645.

Hodgson, E. 1985 Microsomal mono-oxygenases. *Comp. Insect Physiol. Biochem. Physiol.* **11**, 225–321.

Houpt, D. R., Pursey, J. C. & Morton, R. A. 1988 Genes controlling malathion resistance in a laboratory-selected population of *Drosophila melanogaster*. *Genome* **30**, 844–853.

Hung, C. F., Prapaipong, H., Berenbaum, M. R. & Schuler, M. A. 1995 Differential induction of cytochrome P450 transcripts in *Papilio polyxenes* by linear and angular furanocoumarins. *Insect Biochem. Molec. Biol.* **25**, 89–99.

Liu, N. & Scott, J. G. 1996a Genetic analysis of factors controlling activities in LPR house flies, *Musca domestica*. *Biochem. Genet.* **34**, 133–148.

Liu, N. & Scott, J. G. 1996b Genetics of resistance to pyrethroid insecticides in the housefly *Musca domestica*. *Pestic. Biochem. Physiol.* **52**, 116–124.

Liu, N. & Scott, J. G. 1997 Phenobarbital induction in CYP6D1 is due to a trans acting factor on autosome 2 in house flies, *Musca domestica*. *Insect Molec. Biol.* **6**, 77–81.

Maitra, S., Dombrowski, S. M., Waters, L. C. & Ganguly, R. 1996 Three second chromosome-linked clustered Cyp6 genes show differential constitutive and barbital-induced expression in DDT-resistant and susceptible strains of *Drosophila melanogaster*. *Gene* **180**, 165–171.

Megias, A., Saborido, A. & Muncio, A. M. 1984 NADH-cytochrome b5 reductase from the insect *Ceratitis capitata*. Enzyme properties and membrane binding capacity. *Comp. Biochem. Physiol.* B **77**, 679–685.

Nelson, D. R. (and 11 others) 1996 P450 superfamily: update on new sequences, gene mapping, accession numbers, and nomenclature. *Pharmacogenetics* **6**, 1–42.

Omura, T. & Sato, R. 1964 The carbon monoxide-binding pigment of liver microsomes. I. Evidence for its hemoprotein nature. *J. Biol. Chem.* **239**, 2370–2378.

Pittendrigh, B., Aronstein, K., Zinkovsky, E., Andreev, O., Campbell, B., Daly, J., Trowell, S. & ffrench-Constant, R. 1997 Cytochrome P450 genes from *Helicoverpa armigera*: expression in a pyrethroid susceptible and resistant strain. *Insect Biochem. Molec. Biol.* **27**, 507–512.

Plapp, F. W. 1984 The genetic basis of insecticide resistance in the house fly: evidence that a single locus plays a major role in metabolic resistance to insecticides. *Pestic. Biochem. Physiol.* **22**, 194–201.

Rose, R. L., Goh, D., Thompson, D. M., Verma, K. D., Heckel, D. G., Gahan, L. J., Roe, R. M. & Hodgson, E. 1997 Cytochrome P450 (CYP)9A1: the first member of a new CYP family. *Insect Biochem. Molec. Biol.* **27**, 605–615.

Sauphanor, B., Cuany, A., Bouvier, J. C., Brosse, V., Amichot, M. & Bergé, J.-B. 1998 Mechanism of resistance to deltamethrin in field population of *Cydia pomonella* L. (Lepidoptera: Tortricidae). *Pestic. Biochem. Physiol.* **58**, 109–117.

Scott, J. G. & Georghiou, G. P. 1986 Mechanisms responsible for high level of permethrin resistance in LearnPyrR strain of housefly. *Pestic. Sci.* **17**, 195–205.

Scott, J. G., Lee, S. S. T. & Shono, T. 1990 Biochemical changes in the cytochrome P450 monooxygenases of seven insecticide-resistant house fly (*Musca domestica* L.) strains. *Pestic. Biochem. Physiol.* **36**, 127–134.

Scott, J. G., Sridhar, P. & Liu, N. 1996 Adult specific expression and induction of cytochrome P-450lpr in house flies. *Archs Insect Biochem. Physiol.* **31**, 313–323.

Taylor, M. & Feyereisen, R. 1996 Molecular biology and evolution of resistance to toxicants. *Molec. Biol. Evol.* **13**, 719–734.

Vincent, D. R., Moldenke, A. F., Farnsworth, D. E. & Terriere, L. C. 1985 Cytochrome P450 in insects. 6. Age dependency and phenobarbital induction of cytochrome P450, P450 reductase, and monooxygenase activities in susceptible and resistant strains of *Musca domestica*. *Pestic. Biochem. Physiol.* **23**, 171–179.

Waters, L. C. & Nix, C. E. 1988 Regulation of insecticide resistance-related cytochrome P450 expression in *Drosophila melanogaster*. *Pestic. Biochem. Physiol.* **30**, 214–227.

Waters, L. C., Zelhof, A. C., Shaw, B. J. & Chang, L. Y. 1992 Possible involvement of the long terminal repeat of transposable element 17.6 in regulating expression of an insecticide resistance-associated P450 gene. *Drosophila. Proc. Natn. Acad. Sci. USA* **89**, 4855–4859.

Wheelock, G. D. & Scott, J. G. 1990 Immunological detection of cytochrome P450 from insecticide resistant and susceptible house flies (*Musca domestica*). *Pestic. Biochem. Physiol.* **38**, 130–139.

Wilkinson, C. F. & Brattsten, L. B. 1972 Microsomal drug-metabolizing enzymes in insects. *Drug Metab. Rev.* **1**, 153–227.

Xiao-Ping, W. & Hobbs, A. A. 1995 Isolation and sequence analysis of a cDNA clone for a pyrethroid inducible cytochrome P450 from *Helicoverpa armigera*. *Insect Biochem. Molec. Biol.* **25**, 1001–1009.

Zhang, M. & Scott, J. G. 1996 Purification and characterization of cytochrome *b*5 reductase from the house fly, *Musca domestica*. *Comp. Biochem. Physiol.* B **113**, 175–183.

An overview of the evolution of overproduced esterases in the mosquito *Culex pipiens*

Michel Raymond*, Christine Chevillon, Thomas Guillemaud, Thomas Lenormand and Nicole Pasteur

Institut des Sciences de l'Evolution, Laboratoire Génétique et Environnement (C.C. 065), UMR CNRS 5554, Université de Montpellier II, F-34095 Montpellier, France

Insecticide resistance genes have developed in a wide variety of insects in response to heavy chemical application. Few of these examples of adaptation in response to rapid environmental change have been studied both at the population level and at the gene level. One of these is the evolution of the over-produced esterases that are involved in resistance to organophosphate insecticides in the mosquito *Culex pipiens*. At the gene level, two genetic mechanisms are involved in esterase overproduction, namely gene amplification and gene regulation. At the population level, the co-occurrence of the same amplified allele in distinct geographic areas is best explained by the importance of passive transportation at the world-wide scale. The long-term monitoring of a population of mosquitoes in southern France has enabled a detailed study to be made of the evolution of resistance genes on a local scale, and has shown that a resis-tance gene with a lower cost has replaced a former resistance allele with a higher cost.

Keywords: insecticide resistance; selection; adaptation; fitness cost

1. INTRODUCTION

The mosquito *Culex pipiens*, common in temperate and tropical countries, is subjected to insecticide control in many places. Worldwide surveys of resistance to organo-phosphate (OP) insecticides have disclosed that only three loci have developed major resistance alleles (Pasteur *et al.* 1981; Wirth *et al.* 1990; Georghiou 1992; Poirié *et al.* 1992). The first two loci, *Est-2* (or *esterase B*) and *Est-3* (or *esterase A*), code for detoxifying carboxylester hydrolases, and resistant alleles correspond to an esterase overproduction (Fournier *et al.* 1987; Mouchès *et al.* 1987; Poirié *et al.* 1992). Six distinct overproduced allozymes have been described so far at the esterase B locus (B1, B2, B4, B5, B6 and B7) and four (A1, A2, A4 and A5) at the esterase A locus (Pasteur *et al.* 1981, 1984; Raymond *et al.* 1989; Georghiou 1992; Poirié *et al.* 1992; Xu *et al.* 1994; Vaughan & Hemingway 1995). The two esterase loci are closely linked, and are separated by an intergenic DNA fragment of 2–6 kilobases (Rooker *et al.* 1996; Guillemaud *et al.* 1997). The third locus, *Ace.1*, codes the acetylcholin-esterase (insecticide target), and insensitive alleles have been reported in various locations (Raymond *et al.* 1986; Bourguet *et al.* 1996, 1997), although it is not known how many resistant *Ace.1* alleles have occurred independently.

2. DISSECTING MUTATIONS

The overproduction of esterase is the result of two non-exclusive mechanisms. The first mechanism involves gene amplification either of the esterase B locus or, in some situations, of both the esterase A and B loci (Mouchès *et al.* 1986; Raymond *et al.* 1989; Poirié *et al.* 1992; Guillemaud *et al.* 1997; Vaughan *et al.* 1997). The latter situation, the co-amplification of two esterase loci, explains the tight statistical association of some electro-morphs, such as A2 and B2 (Rooker *et al.* 1996; Guillemaud *et al.* 1996). Although, strictly speaking, only A4, A2 and A1 are alleles of the A esterase locus, and only B2 and B4 are alleles at the B esterase locus, A1, A4–B4 and A2–B2 behave as alleles of a single super-gene, which is the result of the complete linkage disequili-brium between the A and B esterase genes produced by the amplification.

The second mechanism, gene regulation, explains the overproduction of esterase A1 (Rooker *et al.* 1996). However, it might also contribute to the overproduction of other variants in addition to gene amplification.

The level of gene amplification varies between the various amplified alleles: for B1, it could reach 250 copies (Mouchès *et al.* 1986), whereas for B4 it has never been found above a few copies (Poirié *et al.* 1992; Guillemaud *et al.* 1997). It varies also within and between populations for a given amplified allele, as shown for example for the A2–B2 amplified allele (Callaghan *et al.* 1998).

3. EVOLUTION AT THE WORLDWIDE SCALE

Overproduced allozymes that are electrophoretically identical are often found in distinct geographic areas. This is the case, for example, for the pair A2–B2, which is present in Africa, South and North America, Asia and Europe, or for B1, present in China and North America. This situation could be explained by a recurrent mutation

*Author for correspondence (raymond@isem.univ-montp2.fr).

process that generates each resistance allele. The other possibility is that each overproduced allele is the result of a non-recurrent mutation, and has subsequently spread within populations owing to its advantage in OP-treated areas, and between populations by migration (not excluding passive transportation). There is now considerable molecular evidence for the latter possibility.

First, restriction maps of the DNA within and around the esterase B structural gene can be built, in susceptible mosquitoes, with a single copy of the gene, as well as in mosquitoes with an amplified esterase gene (the amplicon is larger than the DNA area mapped). When such restriction maps are compared, large differences are observed. For example, two maps from susceptible mosquitoes from the same breeding site have only 40% of their restriction sites in common (Raymond *et al.* 1996). However, when strains with the B2 electromorph are compared, restriction maps are strictly identical, independently of the geographic origins of the insects considered (Raymond *et al.* 1991). A similar situation is found for the B1 electromorph: mosquitoes from various localities within the Americas and China possess the same restriction map (Qiao & Raymond 1995). The similarity of the restriction maps of all B1 (or all B2) haplotypes from diverse and distant geographic areas indicates that all B1 (or all B2) alleles are identical by descent. The same result has been obtained at the sequence level on the intron of the esterase A gene from various geographic origins: the polymorphism found in alleles of non-overproduced esterases is one of the largest thus far described, and all A2 alleles display exactly the same sequence (Guillemaud *et al.* 1996). All of these results could be explained only if a unique molecular event has generated the A2–B2 amplification, which has subsequently spread by migration worldwide. Another independent event is responsible for the B1 amplification, again followed by an extensive migration. It seems that there is a similar situation for the other variants with a large geographical distribution, such as A5–B5 and A4–B4 in the Mediterranean area (Raymond & Marquine 1994; Chevillon *et al.* 1995; Severini *et al.* 1997).

In addition, there is direct (Highton & Van Someren 1970) and indirect (Chevillon *et al.* 1995; Pasteur *et al.* 1995) evidence of large-scale migration of this mosquito by passive transportation by man, and the presence of one female with A2–B2 in an aircraft has been established (Curtis & White 1984). The local invasion of A2–B2 in southern France has been documented: A2–B2 was first found near the international Marseille airport and seaport, and has spread within a few years in all surrounding OP-treated areas (Rivet *et al.* 1993).

The number of independent amplification events at the esterase A and B loci cannot be estimated by just counting the number of overproduced electromorphs. This is due to the following reasons. First, A and B loci are amplified simultaneously, as are A2 and B2 (Rooker *et al.* 1996) and also the associated A4 and B4, and A5 and B5 (Guillemaud *et al.* 1997), so that only one amplification event is responsible for the presence of the two electromorphs. Second, an overproduction of esterase is not necessarily the result of gene amplification, as the overproduction of esterase A1 is due to gene regulation (Rooker *et al.* 1996). Third, the same electromorph could

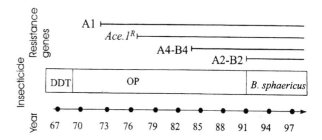

Figure 1. History of insecticide treatments in the Montpellier area, and the occurrence of OP resistance genes. See text for details.

correspond to distinct alleles, as exemplified by B4 and B5 (Poirié *et al.* 1992), and B1 and an unnamed electromorph (Vaughan *et al.* 1995).

Taking into account the protein and DNA studies published to date, the number of independent initial amplification events at both esterase A and B loci is between five and ten. The imprecision arises from the fact that a thorough checking has not been performed yet for all known overproduced esterases. This number corresponds to amplification events that have spread geographically as a result of the advantage they give in insecticide-treated areas, and are therefore at high frequencies and easily detected. A more thorough sampling will probably detect additional events that either are still geographically restricted or are at a low frequency, so that the above estimates should be regarded as a minimum figure. Each known amplification event has spread geographically, sometimes across continents, such as A2–B2 (Raymond *et al.* 1991) and B1 (Qiao & Raymond 1995), and sometimes only in a restricted area, such as A4–B4 in the western Mediterranean (Poirié *et al.* 1992; Severini *et al.* 1997) and A5–B5 in the eastern Mediterranean (Poirié *et al.* 1992; Severini *et al.* 1994, 1997). This relatively low number of independent amplification events, recorded on a world-scale for a pest species with large population sizes, indicates that advantageous mutations (i.e. any molecular event generating a gene amplification at these loci) could be limiting. Once an amplification has occurred, it can apparently spread easily and invade. Clearly, migration cannot be ignored as a driving force in studies and monitoring of insecticide-treated populations of the mosquito *Culex pipiens*.

4. EVOLUTION AT THE LOCAL SCALE

There are very few places in the world where the local evolution of resistance genes has been studied through time. The best-documented place is the Montpellier area (southern France), where information on OP resistance genes has been collected regularly since the occurrence of OP resistance. Different pesticides (DDT, chlorpyrifos, temephos, fenitrothion and *Bacillus sphaericus* toxin) have been used to control *Culex pipiens* (figure 1). The occurrence of OP resistance and resistance genes in this area has already been described in detail elsewhere (Pasteur & Sinègre 1975; Pasteur *et al.* 1981; Raymond *et al.* 1986; Chevillon *et al.* 1995; Guillemaud *et al.* 1998). Briefly, the

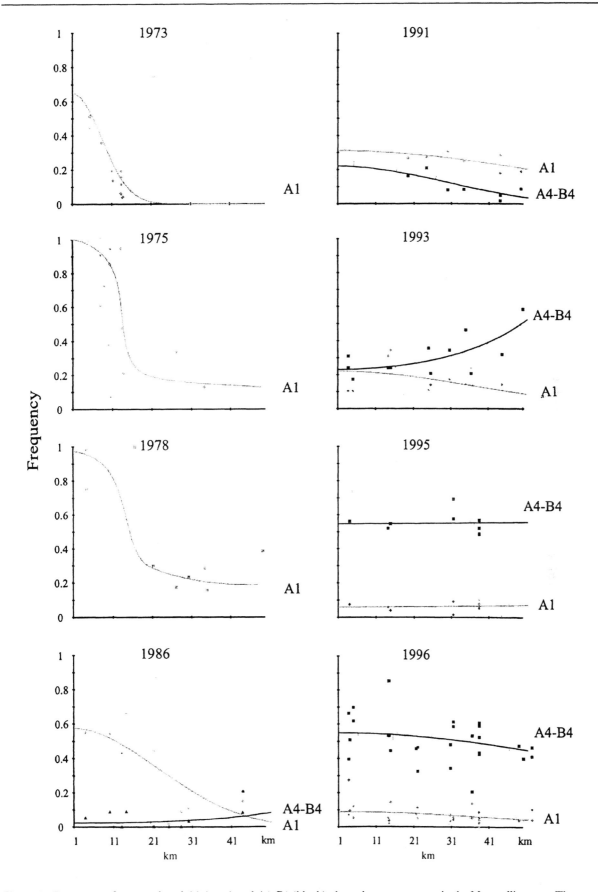

Figure 2. Frequency of overproduced A1 (grey) and A4–B4 (black) along the same transect in the Montpellier area. The frequency of A2–B2 is too low to be represented. The transect starts in the treated area near the sea, crosses the boundary of the treated/non-treated area, and ends 50 km from the sea. Each point represents an independent sample. Eight sampling years are indicated. See Guillemaud et al. (1998) for details.

first resistance gene (overproduced Al esterase that was the result of gene regulation) occurred in 1972, only four years after control began with chlorpyrifos (an OP insecticide). It was followed by the occurrence of a modified target (an insensitive acetylcholinesterase $Ace.1^R$) in 1977, and by two pairs of overproduced esterase A and B allozymes (both the result of gene amplification): A4–B4 in 1984 and A2–B2 in 1991 (figure 1). One particularly convenient feature of *Culex pipiens* in the south of France is that its control has been limited to the populations along the Mediterranean coast. Thus, it is possible to identify, in a linear transect orthogonal to the coastline, an insecticide-treated area (close to the sea) and a non-treated one (further northwest) (figure 2). On this transect, the evolution of overproduced esterases is apparent through time (Guillemaud *et al.* 1998). First, Al increased in frequency, until 1978. It displayed a steep and stable cline, indicating that this allele is associated with a cost, i.e. it is selected against in the non-treated area. At this time, the insensitive $Ace.1^R$ allele occurred, with the global effect of decreasing the frequency of overproduced esterase Al in the treated area. In 1986, a new overproduced esterase allele occurred: A4–B4. Since its first detection, and during its continuous spread, this allele failed to display a typical cline pattern, with a higher frequency in the treated area. This indicates that it is associated with a lower cost than its competing allele Al. It is apparent that an allele replacement has occurred, the first resistance allele Al being replaced over a ten-year period (compare 1986 and 1996 in figure 2) by the less costly A4–B4 allele (Guillemaud *et al.* 1998).

5. CONCLUSION

A handful of overproduced alleles have occurred in *Culex* mosquitoes in response to OP selection throughout the world. Owing to the advantage they provide in OP-treated areas, they have subsequently spread within populations, and then between populations. The latter phenomenon is considerably facilitated by the fact that most OP-treated areas are connected by plane or by other transportation systems that are suitable for passive migration by mosquitoes. As a result, the various overproduced esterase alleles, occurring independently in distinct geographic areas, have come into contact in the same populations, and have competed. Depending of the resistance they provide, the associated cost they possess and various other genetic characteristics (e.g. dominance relationships), one particular allele might eventually increase in frequency and replace the other resistance genes at the same locus. This is illustrated at the local-scale by the replacement of Al by A4–B4 in the Montpellier area, and by the invasion of A2–B2 on a worldwide basis. Which allele will remain (and at which amplification level) is still uncertain, although it seems apparent that selection is sorting alleles with a minimum fitness cost and with a low or intermediate OP resistance.

This work was financed in part by GDR 1105 du programme Environnement, Vie et Société du CNRS, a grant from the Région Languedoc-Roussillon (no. 963223) and by the Entente Interdépartementale pour la Démoustication de Languedoc Roussillon (no. 95.162). T.L. was supported by an ASC from INRA, and T.G. benefited from a MESR fellowship (no. 94137). This is contribution ISEM 98.080 of the Institut des Sciences de l'Evolution (UMR CNRS 5554).

REFERENCES

Bourguet, D., Capela, R. & Raymond, M. 1996 An insensitive acetylcholinesterase in *Culex pipiens* L. mosquitoes from Portugal. *J. Econ. Entomol.* **89**, 1060–1066.

Bourguet, D., Lenormand, T., Guillemaud, T., Marcel, V. & Raymond, M. 1997 Variation of dominance of newly arisen adaptive genes. *Genetics* **147**, 1225–1234.

Callaghan, A., Guillemaud, T., Makate, N. & Raymond, M. 1998 Polymorphism and fluctuations in copy number of amplified esterase genes in *Culex pipiens* mosquitoes. *Insect Molec. Biol.* **7**, 295–300.

Chevillon, C., Pasteur, N., Marquine, M., Heyse, D. & Raymond, M. 1995 Population structure and dynamics of selected genes in the mosquito *Culex pipiens*. *Evolution* **49**, 997–1007.

Curtis, C. F. & White, G. B. 1984 *Plasmodium falciparum* transmission in England: entomological data relative to cases in 1983. *J. Trop. Med. Hyg.* **87**, 101–194.

Fournier, D., Bride, J.-M., Mouchès, C., Raymond, M., Magnin, M., Bergé, J.-B., Pasteur, N. & Georghiou, G. P. 1987 Biochemical characterization of the esterase Al and Bl associated with organophosphate resistance in the *Culex pipiens* L. complex. *Pestic. Biochem. Physiol.* **27**, 211–217.

Georghiou, G. P. 1992 World distribution of esterases involved in organophosphate insecticide resistance in *Culex* mosquitoes, and methods for detection. In *Insecticides: mechanism of action and resistance* (ed. D. Otto & B. Weber), pp. 407–408. Andover: Intercept.

Guillemaud, T., Rooker, S., Pasteur, N. & Raymond, M. 1996 Testing the unique amplification event and the worldwide migration hypothesis of insecticide resistance genes with sequence data. *Heredity* **77**, 535–543.

Guillemaud, T., Makate, N., Raymond, M., Hirst, B. & Callaghan, A. 1997 Esterase gene amplification in *Culex pipiens*. *Insect Molec. Biol.* **6**, 319–327.

Guillemaud, T., Lenormand, T., Bourguet, D., Chevillon, C., Pasteur, N. & Raymond, M. 1998 Evolution of resistance in *Culex pipiens*: allele replacement and changing environment. *Evolution* **52**, 430–440.

Highton, R. B. & Van Someren, C. C. 1970 The transportation of mosquitoes between international airports. *Bull. Wld Hlth Org.* **42**, 334–335.

Mouchès, C., Pasteur, N., Bergé, J. B., Hyrien, O., Raymond, M., Robert de Saint Vincent, B., De Silvestri, M. & Georghiou, G. P. 1986 Amplification of an esterase gene is responsible for insecticide resistance in a California *Culex* mosquito. *Science* **233**, 778–780.

Mouchès, C., Magnin, M., Bergé, J.-B., De Silvestri, M., Beyssat, V., Pasteur, N. & Georghiou, G. P. 1987 Overproduction of detoxifying esterases in organophosphate-resistant *Culex* mosquitoes and their presence in other insects. *Proc. Natn. Acad. Sci. USA* **84**, 2113–2116.

Pasteur, N. & Sinègre, G. 1975 Esterase polymorphism and sensitivity to Dursban organophosphate insecticide in *Culex pipiens pipiens* populations. *Biochem. Genet.* **13**, 789–803.

Pasteur, N., Iseki, A. & Georghiou, G. P. 1981 Genetic and biochemical studies of the highly active esterases A′ and B associated with organophosphate resistance in mosquitoes of the *Culex pipiens* complex. *Biochem. Genet.* **19**, 909–919.

Pasteur, N., Georghiou, G. P. & Iseki, A. 1984 Variation in organophosphate resistance and esterase activity in *Culex quinquefasciatus* Say from California. *Génét. Sélect. Evol.* **16**, 271–284.

Pasteur, N., Marquine, M., Rousset, F., Failloux, A.-B., Chevillon, C. & Raymond, M. 1995 The role of passive migration in the dispersal of resistance genes in *Culex pipiens quinquefasciatus* within French Polynesia. *Genet. Res.* **66**, 139–146.

Poirié, M., Raymond, M. & Pasteur, M. 1992 Identification of two distinct amplifications of the esterase B locus in *Culex pipiens* (L.) mosquitoes from Mediterranean countries. *Biochem. Genet.* **30**, 13–26.

Qiao, C.-L. & Raymond, M. 1995 The same esterase B1 haplotype is amplified in insecticide resistant mosquitoes of the *Culex pipiens* complex from the Americas and China. *Heredity* **74**, 339–345.

Raymond, M. & Marquine, M. 1994 Evolution of insecticide resistance in *Culex pipiens* populations: the Corsican paradox. *J. Evol. Biol.* **7**, 315–337.

Raymond, M., Fournier, D., Bride, J.-M., Cuany, A., Bergé, J., Magnin, M. & Pasteur, N. 1986 Identification of resistance mechanisms in *Culex pipiens* (Diptera: Culicidae) from southern France: insensitive acetylcholinesterase and detoxifying oxidases. *J. Econ. Entomol.* **79**, 1452–1458.

Raymond, M., Beyssat-Arnaouty, V., Sivasubramanian, N., Mouchès, C., Georghiou, G. P. & Pasteur, N. 1989 Amplification of various esterase B's responsible for organophosphate resistance in *Culex* mosquitoes. *Biochem. Genet.* **27**, 417–423.

Raymond, M., Callaghan, A., Fort, P. & Pasteur, N. 1991 Worldwide migration of amplified insecticide resistance genes in mosquitoes. *Nature* **350**, 151–153.

Raymond, M., Qiao, C. L. & Callaghan, A. 1996 Esterase polymorphism in insecticide susceptible populations of the mosquito *Culex pipiens*. *Genet. Res.* **67**, 19–26.

Rivet, Y., Marquine, M. & Raymond, M. 1993 French mosquito populations invaded by A2-B2 esterases causing insecticide resistance. *Biol. J. Linn. Soc.* **49**, 249–255.

Rooker, S., Guillemaud, T., Bergé, J., Pasteur, N. & Raymond, M. 1996 Coamplification of esterase A and B genes as a single unit in the mosquito *Culex pipiens*. *Heredity* **77**, 555–561.

Severini, C., Marinucci, M. & Raymond, M. 1994 Insecticide resistance genes in *Culex pipiens* (Diptera: Culicidae) from Italy: esterase B locus at the DNA level. *J. Med. Entomol.* **31**, 496–499.

Severini, C., Romi, R., Marinucci, M., Guillemaud, T. & Raymond, M. 1997 First record of A5-B5 esterases in an organophosphate-resistant field population of *Culex pipiens* from Italy. *Med. Vet. Entomol.* **11**, 123–126.

Vaughan, A. & Hemingway, J. 1995 Mosquito carboxylesterase Esta21 (A2). *J. Biol. Chem.* **270**, 17044–17049.

Vaughan, A., Rodriguez, M. & Hemingway, J. 1995 The independent gene amplification of electrophoretically indistinguishable B esterases from the insecticide-resistant mosquito *Culex quinquefasciatus*. *Biochem. J.* **305**, 651–658.

Vaughan, A., Hawkes, N. & Hemingway, J. 1997 Co-amplification explains linkage disequilibrium of two mosquito esterase genes in insecticide-resistant *Culex quinquefasciatus*. *Biochem. J.* **325**, 359–365.

Wirth, M., Marquine, M., Georghiou, G. P. & Pasteur, N. 1990 Esterase A2 and B2 in *Culex quiquefasciatus* (Diptera: Culicidae): role in organophosphate resistance and linkage. *J. Econ. Entomol.* **27**, 202–206.

Xu, J., Qu, F. & Liu, W. 1994 Diversity of amplified esterase B genes responsible for organophosphate resistance in *Culex quinquefasciatus* from China. *J. Med. Coll. People's Liberation Army* **9**, 20–23.

A genomic approach to understanding *Heliothis* and *Helicoverpa* resistance to chemical and biological insecticides

David G. Heckel[1,2], Linda J. Gahan[1], Joanne C. Daly[2] and Stephen Trowell[2]

[1]*Department of Biological Sciences, Clemson University, Clemson, SC 29634, USA*
[2]*Division of Entomology, Commonwealth Scientific and Industrial Research Organisation, GPO Box 1700, Canberra 2601, Australia*

Genomics is the comparative study of the structure and function of entire genomes. Although the complete sequencing of the genome of any insect pest is far in the future, a genomic approach can be useful in the study of mechanisms of insecticide resistance. We describe this strategy for *Heliothis* and *Helicoverpa*, two of the most destructive genera of pest moths (Lepidoptera) worldwide. Genome-wide linkage mapping provides the location of major and minor resistance genes. Positional cloning identifies novel resistance genes, even when the mechanisms are poorly understood, as with resistance to *Bacillus thuringiensis* toxins. Anchor loci provide the reference points for comparing the genomes and the genetic architecture of resistance mechanisms among related species. Collectively, these tools enable the description of the evolutionary response of related, but independent, genomes to the common selective pressure of insecticides in the environment. They also provide information that is useful for targeted management of specific resistance genes, and may even speed the search for families of novel insecticidal targets in Lepidoptera.

Keywords: insect genomics; linkage mapping; positional cloning; gene families

1. INTRODUCTION

During the past ten years, there has been unprecedented growth in one of the most paradoxical of all the biological sciences: genomics. In the sense that genomics aims at the comparative study of the structure and function of entire genomes, it is breathtakingly holistic. By analysing those genomes uniformly down to the resolution of single base pairs, it verges on being mindlessly reductionist. The tally of organisms whose genomes have been entirely sequenced has started an exponential growth phase, and biology will never be the same.

The current obsession with DNA sequences often obscures the fact that genomics is made possible by the convergence of two lines of research: genetic and physical. The genetic path began with the invention of linkage maps by Sturtevant and culminates in high-density maps saturated with molecular markers and constructed with data from crosses. The physical path started with Sutton's and Boveri's realization that chromosomes carried the units of heredity and ends in physical maps, which are collections of overlapping fragments of DNA, reassembled back into the same positions they naturally occupy along the chromosome. Points of correspondence between the genetic and physical maps are identified using the DNA sequences of the molecular markers. The two maps merge into one at the instant the entire DNA sequence (all possible markers) is known.

For the genomics of insect pests, this instant is far in the future. In the meantime, there are still aspects of a whole-genome approach that can inform our studies of mechanisms of insecticide resistance. Most useful at present are applications of linkage maps. Our development of a map for the tobacco budworm moth, *Heliothis virescens* (Lepidoptera: Noctuidae) was motivated by the need to analyse multiple resistance mechanisms to chemical insecticides. Currently, about 350 marker loci have been assigned to 31 linkage groups. These markers include 18 allozyme loci resolved by starch gel electrophoresis and enzyme-specific staining, ten restriction-fragment length polymorphisms (RFLPs) defined by anonymous single-copy genomic clones, nine RFLPs defined by known genes, 110 rapid amplified polymorphic DNA (RAPD) markers (Williams *et al.* 1990), and 206 amplified-fragment length polymorphism (AFLP) markers (Vos *et al.* 1995).

This linkage map has been used to determine the number and identity of linkage groups contributing to resistance to a particular insecticide in a particular strain (Heckel *et al.* 1997*b*). Linkage groups are identified using specific marker loci previously localized to them. Resistance can be measured directly by bioassay or indirectly by measuring some property of a resistance mechanism. Quantitative rather than qualitative measures of resistance can provide the extra information necessary to distinguish between resistance genes of major versus minor effect. Depending on the resolution of the mapping study, it may be possible to further discern the number of resistance genes within a linkage group.

We have recently initiated parallel genetic studies in the bollworm *Helicoverpa armigera*. We anticipate that comparison of these two species' maps will shed light on conservation of chromosome structure as well as shared resistance mechanisms: independent 'discoveries' of a common evolutionary solution to the same environmental stress. This comparative genomic approach is well established in mammalian genetics (especially human and mouse) and in the genetics of graminaceous crop plants (rice, wheat, sorghum, maize, etc.). It is starting to show its use in human pathogens, disease vectors and domesticated animals. It is our thesis that a genomic approach also has distinct benefits for the study of insect pests of crops, particularly in understanding and circumventing insecticide resistance, as well as in the discovery of novel insecticidal targets. Heliothine moths, which include some of the most destructive crop-pest species worldwide, offer a tractable and rewarding system for the application of a genome-wide perspective, three aspects of which are reviewed here.

2. GENOME SCANNING FOR RESISTANCE GENES

In our first example, we illustrate how a whole-genome approach adds increased resolution to the genetic characterization of pyrethroid resistance in *H. armigera*. In the standard approach to studying the genetic basis of resistance in a given strain, it is crossed to a susceptible strain, and the hybrids backcrossed to one or both parental strains. The dose–mortality curve of the hybrids is compared to the parental strains to estimate the degree of dominance. The hybrids, which are heterozygous at all resistance loci, transmit resistant and susceptible alleles in the ratio 1:1 to their progeny. The dose–mortality curve of these backcross progeny is examined for goodness-of-fit to a single-locus model. If the hypothesis is not rejected, it is usually concluded that resistance is monogenic; if the goodness-of-fit is poor, the data may be consistent with two or more loci affecting resistance. Although the analysis usually stops there, in certain cases the variance of the tolerance distribution in the F_2 and other generations can be used to estimate the number of effective factors, i.e. the number of loci affecting the trait if they all have the same effect and act additively.

This standard approach has the advantages that it is simple, well-accepted by most researchers, and can provide convincing evidence for a single major resistance locus with a large effect if one exists. Disadvantages include low resolution in discriminating among different multigenic hypotheses within the same strain, and the inability to distinguish between different mechanisms in different strains unless they differ greatly in the magnitude of resistance conferred. A common abuse in practice, although not a disadvantage of the method *per se*, is to use the presence of any sort of inflection in the backcross dose–mortality curve to conclude that a major resistance locus is present, even if the monogenic hypothesis is rejected by goodness-of-fit. This conclusion can obscure the evidence for more than one resistance mechanism and may discourage further exploration to distinguish between two major genes of equal effect, as opposed to distinguishing between one major and several minor genes.

A useful complement to this standard approach uses the ability to score DNA or protein variation at the resistance loci themselves, or at marker loci linked to

them. The segregation pattern of these loci in backcrosses can be compared with the resistance levels of the progeny at selected doses of the insecticide. Often a single dose proves to be diagnostic: a good correlation is found between survivorship at that dose, and possession of the 'resistant' genotype at the marker locus under study. This finding increases the resolution of the genetic analysis because the proportionate contribution of the major locus to the overall level of resistance can be quantified. Two examples from *H. virescens* in which the major locus accounted for some, but not all, of the resistance are the voltage-gated sodium channel (pyrethroid resistance; Taylor *et al.* 1993*a*) and acetylcholinesterase (organophosphate and carbamate resistance; Gilbert *et al.* 1996) mechanisms. However, it is rare to have two different markers, each linked to a different resistance mechanism, studied in the same cross, so that the relative contribution of each can be quantified. Moreover, an approach that focuses only on quantifiable variation in known resistance loci is unable to account for genetic variation at other, unknown loci, and to distinguish it from environmental variation.

A logical (although far from trivial) extension of the single- or double-marker approach is to score enough markers in the backcross so that the entire genome is scanned for segregating genes that contribute to the resistance. In most organisms, a saturated linkage map consisting of closely spaced markers would be needed to ensure that no portion of the genome escapes detection. In certain insects where crossing-over is confined to one sex, one or two markers on each chromosome are sufficient to scan the genome. In flies (Diptera) of the genera *Drosophila* and *Musca*, the genome is broken up into a few large chromosomes, so a few well-placed morphological markers suffice, but the resolution is low. In Lepidoptera such as *Heliothis* and *Helicoverpa*, there are many small chromosomes so the scanning resolution is much higher; but the challenge is to find markers for all the chromosomes.

We have applied the genome-scanning approach to the characterization of the mechanism of fenvalerate resistance in the ANO2 strain of *H. armigera*. Previous work (Daly & Fisk 1992) had implicated pyrethroid detoxification by cytochrome P450 as the major mechanism in this strain, as the resistance was semi-dominant and almost completely eliminated by piperonyl butoxide. Bioassays on backcrosses suggested segregation at a single, major locus, but interpretation was complicated by heterogeneity in different generations of the susceptible strain and in different replicate F_1 lines. We were particularly interested in whether we could find stronger evidence for the existence of a major locus, whether the heterogeneity could be explained by additional resistance loci, and if so what their relative contributions were.

Our initial approach was to use markers already mapped in *H. virescens*, one per chromosome, and to score the homologous markers in *H. armigera*. This would enable us to scan the latter genome for pyrethroid resistance genes and simultaneously establish the correspondence between linkage groups in the two species. This approach is still in progress, as described in more detail in § 4. However, we found that it was faster to develop new markers in *H. armigera* first, so that the scanning could be completed independently of the progress in converting markers from one species to another. To succeed, this alternative approach of 'scanning *de novo*' required a method capable of generating a large number of polymorphic markers so that all 31 chromosomes are marked, even if the identity of the chromosomes is unknown. We found the AFLP technique (Vos *et al.* 1995) very useful for this purpose.

Hybrids between the ANO2 strain and a susceptible strain were produced by single-pair crosses. A female hybrid was backcrossed to a susceptible male, and third-instar larvae were treated with 0.2 µg of fenvalerate in 1 µl of acetone. DNA was isolated from killed offspring and survivors, and 260 AFLPs scored using 11 primer combinations. These were found to group into 31 linkage groups, corresponding to the known number of chromosomes in *H. armigera*. No evidence of crossing-over was found within any of these linkage groups, confirming the earlier cytological observation of the absence of chiasmata in oogenesis in this species (Fisk 1989).

Figure 1 shows a preliminary analysis of the contribution of each of these 31 linkage groups to resistance in the backcross. Only group 13 shows an interaction between genotype and survivorship significant at the Bonferroni level $p < 0.0016$ (the nominal level of 5% adjusted for the fact that 31 independent tests were performed). The only offspring surviving the fenvalerate treatment were heterozygous for group 13; all offspring with two copies of this chromosome from the susceptible strain were killed (as were two heterozygotes). Groups 26 and 29 showed interactions that exceeded the single-test level of 5%, but this cannot be due to resistance alleles from ANO2 because the opposite pattern of mortality was seen. The most likely candidate for a minor resistance gene would be group 17, whose interaction is just significant at 5% and is in the predicted direction for resistance. However, because 31 independent tests were performed, a larger sample size would be required to confidently distinguish the minor locus hypothesis from chance.

Two of the ten AFLPs identified on group 13 were also informative in another backcross between an F_1 male and a susceptible female. Male crossing-over between these markers and the resistance locus, as well as a third AFLP identified in that backcross, permits a linkage map to be constructed (figure 1c). The most likely location of the resistance gene, which we have named *FenR1*, is 9 centiMorgans (cM) from the AFLP marker *atcaglc*.

The results of the analysis to date suggest that a single locus does in fact confer most of the metabolic resistance in the ANO2 strain, and suggests that the heterogeneity found in the earlier studies does not have a genetic basis. We are presently trying to identify which *H. virescens* linkage group corresponds to *H. armigera* AFLP group 13, and to compare the location of *FenR1* with the genomic locations of cytochrome P450 genes that are implicated in resistance in this strain (Pittendrigh *et al.* 1997).

3. POSITIONAL CLONING: FROM GENOME TO RESISTANCE GENE

Once a genomic scan has revealed the location of one or more resistance genes, how can their identity be determined? Most resistance genes that have been cloned were first identified using previous knowledge of the insecticide target site (e.g. acetylcholinesterase or sodium channel) or mechanism of detoxification (e.g. esterases, cytochromes P450 or glutathione transferases), rather than their genomic location. An outstanding exception was the cloning of *Rdl*, a γ-aminobutyric acid (GABA)-gated chloride channel in *Drosophila melanogaster* in which a specific amino-acid substitution confers resistance to dieldrin (ffrench-Constant *et al.* 1991). Although the deletion stocks and other powerful tools of *Drosophila* genetics used to identify *Rdl* as a resistance gene cannot be applied to most other insects, the basic principle can: i.e. positional cloning.

Map-based or positional cloning is a strategy for identifying a gene based primarily on the measurement of its phenotypic effect and knowledge of its genomic location. Although early positional cloning efforts, particularly in humans, were enormous, expensive and time-consuming, recent advances have made it a manageable and feasible strategy more applicable to other organisms. These include faster techniques for screening the thousands of loci needed to find flanking markers and easier methods for constructing genomic libraries with very large inserts.

The steps in positional cloning aim towards continually narrowing the interval of DNA in which the gene of interest must lie; first by linkage analysis and second by construction and analysis of physical maps. The first step is to identify the chromosome or chromosome segment containing the gene, by classical linkage analysis. Second, a saturated map of the region is constructed with as many markers as possible, and their order and spacing relative to the gene of interest is determined by additional linkage analysis. Third, a genomic library is constructed in a vector capable of containing very large DNA fragments. Fourth, the library is screened with markers closest to the gene of interest and flanking it on either side to identify large fragments of DNA that are near to, or contain, the gene. A combination of 'chromosome walking' and additional linkage analysis is employed to construct a contig, an overlapping set of clones that spans the region from one flanking marker to other, in which the gene of interest must lie. Fifth, expressed sequences in the contig are identified, i.e. those that are translated and expressed in the cell, as opposed to repetitive, non-coding or 'junk' DNA. Finally, screening of the remaining candidates based on appropriate expression patterns, gene products, sequence analysis, and other evidence, yields the gene of interest (figure 2).

Positional cloning has been used successfully to identify a wide array of genes of diverse structure, function and phenotypic effect: for example, the gene responsible for cystic fibrosis in humans (Rommens *et al.* 1989); *BRCA1*, which increases breast and ovarian cancer susceptibility in humans (Miki *et al.* 1994); *obese*, which affects obesity in mice (Zhang *et al.* 1994); *nude*, which causes the nude mouse phenotype (Segre *et al.* 1995); *RPS2*, which confers resistance to bacterial infection in *Arabidopsis thaliana* (Bent *et al.* 1994); and *ABI3*, which affects responsiveness to the plant hormone abscisic acid in *A. thaliana* (Giraudat *et al.* 1992). In most cases, the genes found have been completely novel, and shed unexpected new light on the biological process under study. Often it seems in

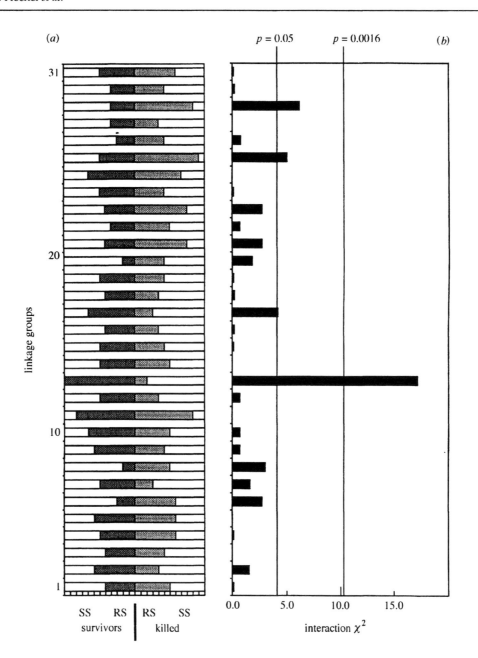

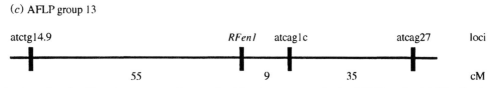

Figure 1. Scanning the *H. armigera* genome for pyrethroid resistance genes in the ANO2 strain. (*a*) Distribution of backcross progeny killed by 0.2 μg of fenvalerate and survivors, for each of the 31 chromosomes identified as AFLP linkage groups. Because there is no crossing-over in the F_1 mother of the backcross progeny, she transmits intact to each of her offspring either the suscep-tible-strain (S) or the resistance-strain (R) homologue of each chromosome. Her mate is from the susceptible strain and always transmits the S homologue. AFLP scores for 12 surviving offspring and 12 killed offspring (*a*) were used to determine the strain origin of both homologues for each linkage group. RS offspring have one chromosome from the resistant strain and one from the susceptible strain (shaded bars); SS have both from the susceptible strain (open bars). At AFLP group 13, for example, all 12 surviving offspring scored were RS and none were SS, whereas two killed offspring were RS and ten were SS. (*b*) Value of the χ^2-statistic testing for an interaction between genotype and mortality. A 2×2 table was constructed for each linkage group and the standard χ^2-test was performed for independence between rows (SS or RS) and columns (killed or survived); $p = 0.0016$ is the Bonferroni confidence level calculated from the nominal 5% level with 31 independent comparisons. Only AFLP group 13 shows a significant departure from independence by this criterion. (*c*) Location of the major resistance locus, *RFen1*, on AFLP group 13. Crossing-over in another cross between an F_1 male and a susceptible female enabled recombination fractions among the AFLP markers and the resistance locus to be measured. Distances are given in centimorgans (cM) using the Haldane mapping function.

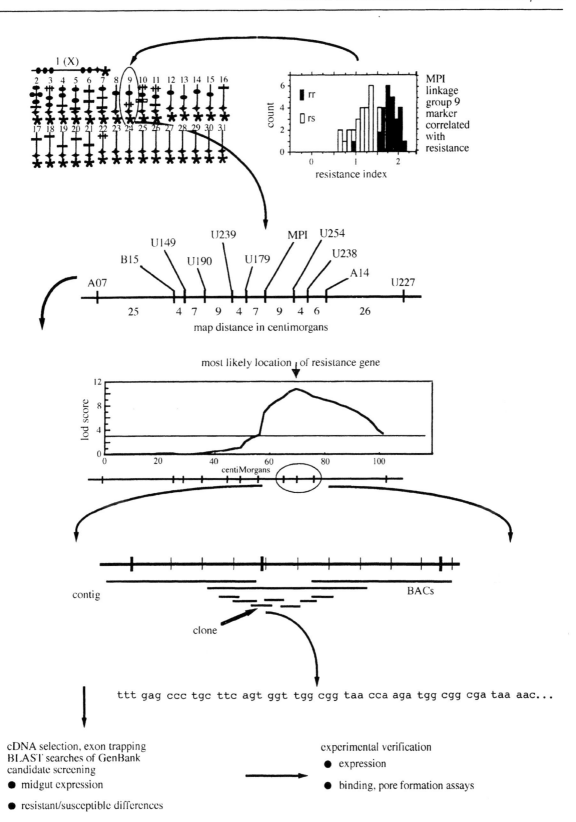

Figure 2. Strategy for positional cloning of the *BtR-4* gene conferring resistance to *Bacillus thuringiensis* Cry1Ac toxin in *H. virescens*. Correlation of the resistance phenotype with the marker locus MPI (mannose phosphate isomerase) enables linkage assignment of *BtR-4* to linkage group 9. Quantitative trait locus (QTL) mapping using ten additional markers on linkage group 9 identifies the most likely location of *BtR-4*. A collection of overlapping bacterial artificial chromosome (BAC) clones is identified that covers the relevant area, and fine-scale and physical mapping of this contig identifies appropriate subclones for sequencing. The sequences are analysed to identify suitable candidates. Finally, *BtR-4* is identified among these candidates by experimental approaches.

retrospect that the gene could not have been cloned in any other way.

A prime candidate for the positional cloning approach is a major gene conferring resistance to *Bacillus thuringiensis* (Bt) crystal protein toxin CrylAc in *Heliothis virescens*. Although Bt resistance has not yet been detected in field populations of *H. virescens*, several strains have been selected in the laboratory. The most resistant strain produced to date is YHD2 (Gould *et al.* 1995), which has attained up to 10 000-fold resistance to CrylAc. Recently, we have shown that nearly 80% of the CrylAc resistance in YHD2 is due to the effects of a single autosomal gene on linkage group 9 of *H. virescens* (Heckel *et al.* 1997a). This gene, *BtR-4*, is the most potent Bt-resistant gene known in any insect. The resistant allele *r* is recessive, thus *rs* heterozygotes are nearly as susceptible as *ss* homozygotes. However, *rr* resistant homozygotes can survive high doses of CrylAc that are lethal to all susceptibles, and can grow and complete development at lower doses that completely inhibit growth of susceptibles. Now that Monsanto's transgenic 'Bollgard' cotton, expressing the CrylAc toxin, is commercially grown in the United States, we can expect to see the resistant allele *r* at *BtR-4* favoured by selection.

The selection response may well be immediate, because the resistant YHD2-type *r* allele is already present at a frequency of *ca.* 10^{-3} in field populations. When *H. virescens* males were captured in the field using pheromone traps and mated to laboratory YHD2 females (all homozogous *rr* for the resistant allele), 3 out of 1025 field-caught males were found to be *rs* heterozygotes based on high CrylAc resistance in their progeny (Gould *et al.* 1997). By conventional expectations, this is an extraordinarily high preselection field frequency for resistance alleles. With this information, it is clearly fortunate that a resistance-management strategy has been implemented from the very first commercial release of Bollgard cotton. Under the terms of the licence agreement with Monsanto, each farmer growing Bollgard must also provide for a 'refuge' where any budworms that develop do so without Bt exposure. The intent is to provide a pool of susceptible insects for mating with the few resistant survivors of selection from the Bollgard crop, to dilute resistance alleles and keep as many of them as possible in the heterozygous state. But how successful will this strategy be, and how long will it delay resistance? No successful resistance-management strategy for tobacco budworm on transgenic cotton can ignore the effects of *BtR-4*. It would be very useful to monitor the frequency of the resistant YHD2-type allele in the field, to determine how well the strategy is working. Yet the only method available at present is the extremely laborious one of conducting thousands of single-pair matings and progeny tests as done by Gould *et al.* (1997). Cloning *BtR-4* would make this much easier by enabling the development of gene-specific markers for field monitoring.

In preparation for positional cloning of *BtR-4*, we have already mapped it to a 10-cM region of linkage group 9 and isolated flanking RAPD markers. In future, we will be working on finer localization with AFLPs, construction of a library in a bacterial artificial chromosome (BAC) vector and assembly into a set of overlapping clones covering the *BtR-4* region, isolating candidate genes from these clones by exon trapping and hybrid selection with complementary DNA, and confirming experimentally that one of these candidates is the *BtR-4* gene (figure 2).

4. ANCHOR LOCI FOR COMPARING RELATED GENOMES

Scanning a genome and positional cloning are activities conducted for a single species. We may want to use a comparative approach to learn more about the occurrence of common resistance mechanisms in different species. How would we perform an interspecific comparison of genome scans? Can we use positional information from one species to assist in cloning a resistance gene from another? Both endeavours require that we have anchor loci to serve as reference points for the comparison.

Ideal anchor loci are present in exactly one copy per haploid genome, and can be easily detected, scored and mapped in all species to be compared. When chromosomal rearrangements have occurred after the divergence of two species from a common ancestor, they will be revealed by the different mutual linkage relationships of the anchors in the two species. In conserved regions of the genome not affected by such chromosomal rearrangements, the position relative to anchors will define the interspecies correspondence between the intervening genes. This approach can be used to provide indirect evidence that resistance genes in two different species are homologous, if they occur in the same relative place in the genomes of both species.

When the resistance genes are already known, of course, there is more direct evidence of homology. The finding that the same mutation in the voltage-gated sodium channel confers pyrethroid resistance in houseflies (Williamson *et al.* 1996), and cockroaches (Miyazaki *et al.* 1996; Dong *et al.* 1997), is a powerful example of this. Within each species, the resistance phenotype is genetically linked to molecular variation at the sodium channel. The resistance locus itself serves as the anchor connecting the results of the two species. It shows that resistance in both is conferred by modification of the same target, and that the target was modified in the same way in both species. It also serves as an anchor in comparisons with *H. armigera* and *H. virescens*, in which different modifications of the sodium channel appear to have occurred (Park & Taylor 1997; Head *et al.* 1998). Gene duplication can complicate the comparisons, even for anchor loci that are also resistance genes. Recent evidence points to the existence of two acetylcholinesterase genes in the mosquito *Culex pipiens*, only one of which appears to be modified to confer resistance to organophosphates (Malcolm *et al.* 1998).

When the resistance genes have not yet been cloned, the indirect approach can be used to obtain evidence for homologous mechanisms in different species. Resistance to Bt toxins in Lepidoptera provides a timely example of the need for anchor loci for this purpose. The homology-based approach is needed because physiological and biochemical comparisons are still inconclusive. Both *Plodia interpunctella* (Van Rie *et al.* 1990) and *Plutella xylostella* (Ferré *et al.* 1991) show a striking correlation between high CrylAc resistance and greatly reduced CrylAc toxin binding to midgut vesicles. Yet it is not known whether the same binding target is involved in the two species. Moreover, *H. virescens* (Lee *et al.* 1995) as well as many other species show a much less convincing relationship between resistance and toxin binding. On the other hand, suggestive evidence for homologous Bt targets among Lepidoptera is provided by the finding that midgut aminopeptidase N has been shown to bind CrylAc toxin in many species (Knight *et al.* 1994; Sangadala *et al.* 1994; Gill *et al.* 1995; Valaitis *et al.* 1995; Luo *et al.* 1997). But even here, there is no evidence to date of homologous resistance mechanisms, i.e. modified aminopeptidases with reduced toxin binding or indeed any effect on resistance levels.

We believe that a comparison of *Heliothis* and *Helicoverpa* genomes will reveal homologous resistance mechanisms. Bt resistance genes have already been shown to occur on five linkage groups in *H. virescens* (Heckel *et al.* 1996). As *H. armigera* develops Bt resistance, we will map those loci as well. For comparison, though, it may be faster to develop anchor loci for the *H. virescens* Bt resistance loci and score them in *H. armigera* strains currently undergoing selection for resistance, than to independently isolate putative Bt resistance genes from *H. armigera* first. More broadly, we plan to compare *H. virescens* to other Lepidoptera that have evolved Bt resistance. Tabashnik *et al.* (this issue) have suggested that some (but not all) resistant strains of *P. xylostella*, *P. interpunctella* and *H. virescens* may have a common genetic mechanism of CrylAc resistance involving a major gene. This hypothesis would be strongly supported if resistance in these strains (but not necessarily others) was shown to be linked to the same anchor locus for all three species.

Among the various classes of marker loci in use, there is unfortunately a trade-off between suitability for genome scanning and suitability as anchors. RAPDs and AFLPs, for which it is easiest to score large numbers of polymorphisms within species, are the hardest to establish with confidence as homologous between species. Yet Lepidoptera with many small chromosomes provide exacting demands for large numbers of markers. Any chromosome that lacks an anchor cannot be compared among species. Increasing densities of anchor loci permit increasingly refined comparisons (figure 3). One anchor locus per chromosome would suffice to detect chromosomal fusions, but most translocations would go undetected. Two anchor loci per chromosome could detect some translocations and provide an overall test of conserved synteny, but would not provide any information about conservation of gene order. Three anchor loci per chromosome would enable coarse-grained tests of conserved order between species, but would miss micro-scale rearrangements. Detection of these might best be accomplished by comparison of physical maps, as described in the next section.

An extensive list of anchor loci has been proposed for mammals (O'Brien *et al.* 1993), but unfortunately most of these seem unlikely to have easily identifiable homologues in other taxa. As a first step in developing a set of anchor loci for genomic comparisons in insects, we have chosen to focus on the gene–enzyme systems studied by population geneticists since the 1960s. These enzymes catalyse

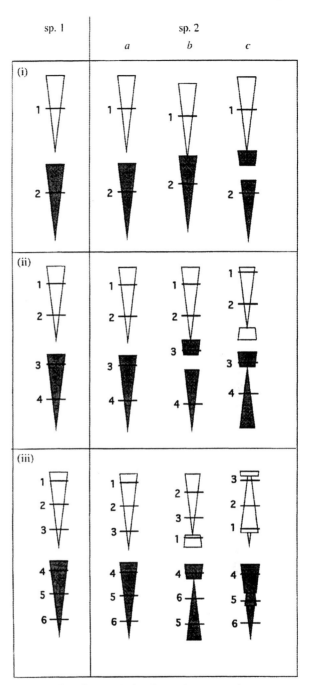

Figure 3. Use of anchor loci to detect chromosomal rearrangements. Two non-homologous chromosomes are depicted as wedges. Anchor loci are shown on the chromosomes of species 1. Scoring the same anchor loci in species 2 will detect some but not all chromosomal rearrangements, depending on the number of anchors per chromosome. In condition (*a*), species 2 has no rearrangements relative to species 1. The anchors shown enable this situation to be distinguished from condition (*b*), but not from condition (*c*). (i) One anchor per chromosome can detect chromosomal fusions, but not certain translocations. (ii) Two anchors per chromosome can detect gross translocations, but not inversions. (iii) Three anchors per chromosome can detect gross inversions, but not micro-inversions.

steps in glycolysis, the citric acid cycle, and nucleotide and amino-acid metabolism that are universal to all living organisms. Variation in electrophoretic mobility of the allozymes—allelic forms of the same enzyme—has

Table 1. *Proposed anchor loci based on allozymes mapped in H. virescens*

TPI	triosephosphate isomerase
MPI	mannose-6-phosphate isomerase
PGM	phosphoglucomutase
G6P	glucose-6-phosphate dehydrogenase
SOD	superoxide dismutase
PGI	phosphoglucose isomerase
Fum	fumarate hydratase
MDHm	malate dehydrogenase (mit)
SoDH	sorbitol dehydrogenase
IDHm	isocitrate dehydrogenase (mit)
AATm	aspartate aminotransferase (mit)
AATc	aspartate aminotransferase (cyt)

been used to map such loci in a wide variety of species. Scoring nucleotide variation in the genes encoding these enzymes would make them informative as anchors even in species where the allozymic variation is rare or absent. With the increasing amount of information from genome sequencing projects, genes for the same enzyme can be identified in several different species and aligned to identify evolutionarily conserved regions. We have designed degenerate polymerase chain reaction (PCR) primers that are complementary to these regions for amplifying a set of genes that have already been mapped in *H. virescens* on the basis of allozymic variation (table 1). We will use these to begin to establish the correspondence between the *H. armigera* linkage groups identified in the genome-scanning for pyrethroid resistance, and the linkage groups of *H. virescens*. We encourage the use of these primers by workers on other insects as well, because the most useful anchors are those that have been scored in a large number of species.

5. TOWARDS A PHYSICAL MAP OF THE HELIOTHINE GENOME

So far, our main approach has been genetic, and our only venture into the physical realm was discussion of a collection of contiguous clones covering a portion of linkage group 9 for positional cloning of *BtR-4*. Extension of this coverage to all the chromosomes would constitute a first-generation physical map. Technological advances have brought this step within reach. We have thus begun to investigate the benefits a physical map would bring, and the feasibility of constructing one in *Heliothis* or *Helicoverpa*.

H. virescens has a genome size of about 403 million base pairs (Taylor *et al.* 1993*b*). A BAC library with 60 000 clones containing inserts of 100–150 kilobases would provide about a 20-fold coverage of the genome. Colonies of all clones can be spotted onto filters in a high-density array for efficient detection by hybridization. With every region of the genome covered by multiple overlapping BAC inserts, fingerprinting techniques based on matching restriction fragment profiles of the clones can be used to piece them together into locally contiguous groups that correspond to different chromosomal regions. Hybridization of genetically mapped AFLP fragments to the array would provide the correspondence between the genetic map and the physical map.

Such a physical map would offer both short-term and long-term benefits. In the short term, it would speed up mapping of newly cloned genes, by replacing RFLP linkage mapping with hybridization to the BAC filters. The whole-genome coverage would facilitate faster and more complete characterization of tandemly repeated gene families involved in resistance, such as cytochromes P450 and esterases. It would speed up positional cloning of novel resistance genes by providing immediate access to clones covering the chromosomal area harbouring the resistance gene, rather than having to 'walk' *de novo* across the region spanned by the flanking markers each time.

In the longer term, a physical map of BAC clones permits a very efficient strategy for large-scale sequencing, whether the whole genome is being contemplated or attention is limited to areas with multigene families. BAC libraries have already been constructed for the silkworm *Bombyx mori*, another Lepidopteran. Comparison of the bombycid and heliothine physical maps would provide the first detailed information on genome rearrangement within Lepidoptera at a smaller-scale than linkage mapping with anchor loci can detect. Comparing both maps to the physical map of *Drosophila melanogaster* will reveal whether any gene collinearity has been preserved at the fine-scale, as a vestige of the gene order of their common insect ancestor.

The degree of collinearity found will indicate how feasible it will be to directly clone novel insecticidal targets using conserved synteny. This could bring discoveries in insect model systems directly to bear on biorational insecticide discovery. *Bombyx mori*, which has been studied intensively on the genetic, physiological and biochemical levels, has been proposed as a model Lepidopteran (Goldsmith 1995). *Drosophila melanogaster* is currently being used to elucidate evolutionarily conserved signalling pathways as targets for novel drugs in the pharmacogenetic approach (Scangos 1997). The *Heliothis* homologues of genes discovered in either model system could be cloned by synteny, providing the fraction of the genome collinear in all three species is high. Although this fraction is currently unknown for insects, extensive gene collinearity has been discovered within mammals (e.g. human and mouse, where it has been used to assist positional cloning) and within crop plants of the grass family (e.g. rice, wheat, sorghum and maize).

6. CONCLUSION

In closing, we emphasize two features of the genomics approach that are hard to quantify but extremely important. First, and most elusive, is the property of 'completeness' or 'exhaustiveness' that is just manifesting itself in the study of the dozen or so organism whose genomes have been entirely sequenced. Not only do we know all the genes such an organism has, we also know which genes it does not have. If genes crucial for a particular biochemical pathway are absent from the genome, we know that pathway is not used by that organism, and if the result of that pathway is achieved, it must be done some other way. Second, is the paramount nature of the comparative approach. There is now far more information on structural genomics (i.e. DNA sequences) than on functional genomics (the biological function of genes).

The vast majority of sequenced genes resulting from the genome projects have not been studied directly to ascertain their function. Instead, their function is inferred on the basis of sequence similarity to genes whose function has been experimentally determined. If circularity can be avoided, the combination of completeness and globally applied sequence comparisons provides a powerful new tool to study how the evolutionary process has resulted in the current diversity of life from a single common ancestor. It is a tool we must learn to use, and to apply to problems of practical importance such as understanding and circumventing insecticide resistance—the multiple 'solutions' produced by evolution in insect pests to the selective challenge of pesticides in their environment.

Our research on *Heliothis* and *Helicoverpa* has been supported by the USDA National Research Initiative Competitive Grants Program, the USDA Biotechnology Risk Assessment Program, the Bt Management Working Group, an NSF-EPSCoR grant to the State of South Carolina, the Australian Cotton Research and Development Corporation, and the Australian Cooperative Research Centre for Sustainable Cotton Production. D.G.H. was the recipient of a Fulbright Senior Scholar Award.

REFERENCES

Bent, A. F., Kunkel, B. N., Dahlbeck, D., Brown, K. L., Schmidt, R., Giraudat, J., Leun, J. & Staskawicz, B. J. 1994 *RPS2* of *Arabidopsis thaliana*: a leucine-rich repeat class of plant disease resistance genes. *Science* **265**, 1856–1860.

Daly, J. C. & Fisk, J. H. 1992 Inheritance of metabolic resistance to synthetic pyrethroids in Australian *Helicoverpa armigera* (Lepidoptera: Noctuidae). *Bull. Entomol. Res.* **82**, 5–12.

Dong, K. E. 1997 A single amino acid change in the para sodium channel protein is associated with knockdown-resistance (*Kdr*) to pyrethroid insecticides in German cockroach. *Insect Biochem. Molec. Biol.* **27**, 93–100.

Ferré, J., Real, M. D., Van Rie, J., Jansens, S. & Peferoen, M. 1991 Resistance to the *Bacillus thuringiensis* bioinsecticide in a field population of *Plutella xylostella* is due to a change in a midgut membrane receptor. *Proc. Natn. Acad. Sci. USA* **88**, 5119–5123.

ffrench-Constant, R. H., Mortlock, D. P., Schaffer, C. D., McIntyre, R. J. & Roush, R. T. 1991 Molecular cloning and transformation of cyclodiene resistance in *Drosophila*: an invertebrate γ-aminobutyric acid subtype A receptor locus. *Proc. Natn. Acad. Sci. USA* **88**, 7209–7213.

Fisk, J. H. 1989 Karyotype and achiasmatic female meiosis in *Helicoverpa armigera* (Hübner) and *H. punctigera* (Wallengren) (Lepidoptera: Noctuidae). *Genome* **32**, 967–971.

Gilbert, R. D., Bryson, P. K. & Brown, T. M. 1996 Linkage analysis of insecticide-resistant acetylcholinesterase in *Heliothis virescens*. *Biochem. Genet.* **34**, 297–312.

Gill, S. S., Cowles, E. A. & Francis, V. 1995 Identification, isolation and cloning of a *Bacillus thuringiensis* CryIAc toxin-binding protein from the midgut of the lepidopteran insect *Heliothis virescens*. *J. Biol. Chem.* **270**, 27 277–27 282.

Giraudat, J., Hauge, B. M., Valon, C., Smalle, J., Parcy, F. & Goodman, H. M. 1992 Isolation of the *Arabidopsis ABI3* gene by positional cloning. *Pl. Cell* **4**, 1251–1261.

Goldsmith, M. R. 1995 Genetics of the silkworm: revisiting an ancient model system. In *Molecular model systems in the Lepidoptera* (ed. M. R. Goldsmith & A. S. Wilkins), pp. 21–76. Cambridge University Press.

Gould, F., Anderson, A., Reynolds, A., Bumgarner, L. & Moar, M. 1995 Selection and genetic analysis of a *Heliothis*

virescens (Lepidoptera: Noctuidae) strain with high levels of resistance to *Bacillus thuringiensis* toxins. *J. Econ. Entomol.* **88**, 1545–1559.

Gould, F., Anderson, A., Jones, A., Sumerford, D., Heckel, D. G., Lopez, J., Micinski, S., Leonard, R. & Laster, M. 1997 Initial frequency of alleles for resistance to *Bacillus thuringiensis* toxins in field populations of *Heliothis virescens*. *Proc. Natn. Acad. Sci. USA* **94**, 3519–3523.

Head, D. J., McCaffery, A. R. & Callaghan, A. 1998 Novel mutations in the *para*-homologous sodium channel gene associated with phenotypic expression of nerve insensitivity resistance to pyrethroids in heliothine Lepidoptera. *Insect Molec. Biol.* **7**, 191–196.

Heckel, D. G., Gahan, L. C., Gould, F. & Tabashnik, B. E. 1996 Mapping major and minor genes conferring resistance to Bt toxins in Lepidoptera. In *Proceedings of the second Pacific Rim conference on biotechnology of* Bacillus thuringiensis *and its impact on the environment*, 4–8 November 1996, Chiang Mai, Thailand, pp. 468–480. Bangkok: Entomology & Zoology Association of Thailand.

Heckel, D. G., Gahan, L. C., Gould, F. & Anderson, A. 1997*a* Identification of a linkage group with a major effect on resistance to *Bacillus thuringiensis* CrylAc endotoxin in the tobacco budworm (Lepidoptera: Noctuidae). *J. Econ. Entomol.* **90**, 75–86.

Heckel, D. G., Gahan, L. C., Gould, F., Daly, J. C. & Trowell, S. 1997*b* Genetics of *Heliothis* and *Helicoverpa* resistance to chemical insecticides and to *Bacillus thuringiensis*. *Pestic. Sci.* **51**, 251–258.

Knight, P. J. K., Crickmore, N. & Ellar, D. J. 1994 The receptor for *Bacillus thuringiensis* CryIA(c) delta-endotoxin in the brush border membrane of the lepidopteran *Manduca sexta* is aminopeptidase *N*. *Molec. Microbiol.* **11**, 429–436.

Lee, M. K., Rajamohan, R., Gould, F. & Dean, D. H. 1995 Resistance to *Bacillus thuringiensis* CryIA δ-endotoxins in a laboratory-selected *Heliothis virescens* strain is related to receptor alteration. *Appl. Environ. Microbiol.* **61**, 3836–3842.

Luo, K., Tabashnik, B. E. & Adang, M. J. 1997 Binding of *Bacillus thuringiensis* CrylAc toxin to aminopeptidase in susceptible and resistant diamondback moths (*Plutella xylostella*). *Appl. Environ. Microbiol.* **63**, 1024–1027.

Malcolm, C. A., Bourguet, D., Ascolillo, A., Rooker, S. J., Garvey, C. F., Hall, L. M. C., Pasteur, N. & Raymond, M. 1998 A sex-linked *Ace* gene, not linked to insensitive acetylcholinesterase-mediated insecticide resistance in *Culex pipiens*. *Insect Molec. Biol.* **7**, 107–120.

Miki, Y. (and 44 others) 1994 A strong candidate for the breast and ovarian cancer susceptibility gene *BRCA1*. *Science* **266**, 66–71.

Miyazaki, M., Ohyami, K., Dunlap, D. Y. & Matsumura, F. 1996 Cloning and sequencing of the *para*-type sodium channel gene from susceptible and *kdr*-resistant German cockroaches (*Blatella germanica*) and housefly (*Musca domestica*). *Molec. Gen. Genet.* **252**, 61–68.

O'Brien, S. J., Womack, J. E., Lyons, L. A., Moore, K. I., Jenkins, N. A. & Copeland, N. G. 1993 Anchored reference loci for comparative genome mapping in mammals. *Nature Genet.* **3**, 103–112.

Park, Y. & Taylor, M. F. J. 1997 A novel mutation L1029H in sodium channel gene *hscp* associated with pyrethroid resistance for *Heliothis virescens* (Lepidoptera: Noctuidae). *Insect Biochem. Molec. Biol.* **27**, 9–13.

Pittendrigh, B., Aronstein, K., Zinkovsky, E., Andreev, O., Campbell, B., Daly, J., Trowell, S. & ffrench-Constant, R. H. 1997 Cytochrome P450 genes from *Helicoverpa armigera*: expression in a pyrethroid-susceptible and resistant strain. *Insect Biochem. Molec. Biol.* **27**, 507–512.

Rommens, J. M. (and 14 others) 1989 Identification of the cystic fibrosis gene: chromosome walking and jumping. *Science* **245**, 1059–1065.

Sangadala, S., Walters, F. S., English, L. H. & Adang, M. J. 1994 A mixture of *Manduca sexta* aminopeptidase and phosphatase enhances *Bacillus thuringiensis* insecticidal CryIAc toxin binding and ^{86}Rb$^+$–K$^+$ efflux *in vitro*. *J. Biol. Chem.* **269**, 10 088–10 092.

Scangos, G. 1997 Drug discovery in the postgenomic era. *Nat. Biotechnol.* **15**, 1220–1221.

Segre, J. A., Nemhauser, J. L., Taylor, B. A., Nadeau, J. H. & Lander, E. S. 1995 Positional cloning of the *nude* locus: genetic, physical, and transcription maps of the region and mutations in the mouse and rat. *Genomics* **28**, 549–561.

Taylor, M. F. J., Heckel, D. G., Brown, T. M., Kreitman, M. E. & Black, B. 1993*a* Linkage of pyrethroid resistance to a sodium channel locus in the tobacco budworm. *Insect Biochem. Molec. Biol.* **23**, 763–775.

Taylor, M., Zawadski, J., Black, B. & Kreitman, M. 1993*b* Genome size and endopolyploidy in pyrethroid-resistant and susceptible strains of *Heliothis virescens*. *J. Econ. Entomol.* **86**, 1030–1034.

Valaitis, A., Lee, M. K., Rajamohan, F. & Dean, D. H. 1995 Brush border membrane aminopeptidase N in the midgut of the gypsy moth serves as the receptor for the CryIA(c) δ-endotoxin of *Bacillus thuringiensis*. *Insect Biochem. Molec. Biol.* **25**, 1143–1151.

Van Rie, J., McGaughey, W. H., Johnson, D. E., Barnett, B. D. & Van Mellaert, H. 1990 Mechanism of insect resistance to the microbial insecticide *Bacillus thuringiensis*. *Science* **247**, 72–74.

Vos, P. (and 10 others) 1995 AFLP: a new concept for DNA fingerprinting. *Nucl. Acids Res.* **23**, 4407–4414.

Williams, J. G. K., Kubelik, A. R., Livak, K. J., Rafalski, A. & Tingey, S. V. 1990 DNA polymorphisms amplified by arbitrary primers are useful as genetic markers. *Nucl. Acids Res.* **18**, 6531–6535.

Williamson, M. S., Martinez-Torres, D., Hick, C. A. & Devonshire, A. L. 1996 Identification of mutations in the *Drosophila para*-type sodium channel gene associated with knockdown resistance (*kdr*) to pyrethroid insecticides. *Molec. Gen. Genet.* **252**, 51–60.

Zhang, Y., Proenca, R., Maffei, M., Barone, M., Leopold, L. & Friedman, J. M. 1994 Positional cloning of the mouse *obese* gene and its human homologue. *Nature* **372**, 425–432.

The influence of the molecular basis of resistance on insecticide discovery

Michael D. Broadhurst

Zeneca Ag Products, Western Research Center, 1200 South 47th Street, Richmond, CA 94804, USA

This paper focuses on the process of invention and development of new insecticides and the impact of current research in resistance mechanisms on that process. The topic is introduced in the context of (i) the critical need to develop new insect-control agents to ensure a continued supply of high-quality food and fibre; (ii) how resistance development will continue to influence the potential to ensure the supply of these essentials; and (iii) why new insect-control technology is welcomed by growers.

The main section of the paper describes a generic agrochemical invention process and discusses the impact that an understanding of the molecular basis of resistance will have on the various stages of this process, using specific examples to illustrate these points. By focusing on insecticide invention, this paper provides a context in which other information more specific to insecticide resistance from this issue can be understood.

Keywords: resistance; insecticide; invention; discovery; agriculture

1. INTRODUCTION

Crop protection is a research-based business where a number of drivers demand a continual flow of new products. Although reasonably effective products exist in most market segments, new products that allow growers to produce their crop in a safe and economical manner are always welcomed. These new products must conform to high standards of safety with respect to their potential to affect the health of those who use them, the quality of the food we eat and the environment. These factors, together with the need to provide an adequate return to those who invest in the companies that conduct the research and development required to discover new crop-protective agents, impose many diverse scientific and commercial challenges on the invention process.

Development of resistance to existing products is an important driver for new methods to control insects, plant pathogens and weeds in many market segments. This paper focuses on the impact of our increasing understanding of the detailed mechanisms of resistance to insect-control agents at the molecular level on the process of invention of suitable new chemistry.

2. BACKGROUND

One study predicts that the world population will continue its present growth to approximately 11 billion during the period between 2025 and 2050 and then level off (UNFPA, unpublished). Associated with this population growth, the amount of land available to grow food will continue to diminish (UNFPA, unpublished). In addition, individuals are becoming more sophisticated in their tastes and increased wealth is allowing people to make choices, potentiating the other factors by shifting agriculture towards less efficient foods to aggravate the food-supply problems. Similar forces are affecting the supply of natural fibres produced directly or indirectly through agriculture.

There is no doubt that methods of crop protection have been a major contributor in our current ability to produce a relatively abundant and good-quality food and fibre supply. One report from a study of several major crops clearly demonstrates that crop protection has improved yields (Oerke *et al.* 1994). As an example, for rice it was estimated that 65% of crop yield is saved by crop protection. The data also suggest that superior technology could improve yields still further.

These factors together indicate that we will require more high-quality food in the future and superior crop-protection methods to allow us to produce and protect this food.

A number of elements influence our ability to produce high-quality crops in good yield. These include inputs such as crop protection, fertilizers, irrigation, and seed improved through breeding or genetic manipulation. Equally important are less controllable determinants such as soil erosion, limitations on the use of pesticides through regulations, and pest resistance. Although I will focus in this paper on the two elements of pest resistance and crop protection, it is difficult to separate these from the many other important influences, some of which were mentioned above. Therefore, this discussion will include a variety of other factors that are important to the invention process for new crop-protection methods.

There is no doubt that pest resistance can have devastating implications for a grower's ability to produce a crop, in some instances leading to complete crop loss. Such instances of crop loss occurred in cotton during 1995, one of the worst years in history for the control of tobacco budworm (*Heliothis virescens*), in the southeast United States. In some areas, owing to the presence of resistant individuals, no insecticide or combination of insecticides could control damage. Two years earlier, a study from Louisiana State University of eight field-collected populations of tobacco budworm showed high levels of tolerance to four different insecticides representing three different modes of action (Graves *et al.* 1993).

Fortunately, in the two subsequent growing seasons, tobacco budworm pressure has been lower. Furthermore, we now have access to several new tools with which to fight this pest. Notable in this regard was the introduction of Bollgard® cotton. This technology confers crop resistance to some cotton pests via the expression of the Bt (*Bacillus thuringiensis*) endotoxin in the exposed parts of the cotton plant. Since its introduction, Bollgard® cotton appears to have provided excellent control of tobacco budworm in practical applications under light to moderate insect pressure in most growing areas.

Nevertheless, the Bt endotoxin is intrinsically less potent against the related cotton bollworm (*Heliothis zea*) and this appears to have led to some instances of bollworm damage in cotton not otherwise treated. This example illustrates a key point. No single insecticide or other technology is the sole answer to the grower's insect-control problems. This was confirmed in large-field plots where overspraying Bt Cotton with the synthetic pyrethroid insecticide λ-cyhalothrin provided significant yield increases and consequent economic benefits to the growers (Mink *et al.* 1997). Although firm evidence regarding the origin of the observed yield improvements is lacking, one can reasonably speculate that these arose as a result of factors such as control of pests that were not susceptible to the Bt endotoxin or repellency to bollworm moths.

A second study carried out in the laboratory demonstrated that insects that survived exposure to sprayable Bt products at generally sub-lethal doses were more susceptible to λ-cyhalothrin than unexposed pests (Harris *et al.* 1998). This study concluded that oversprays of λ-cyhalothrin on Bt Cotton would tend to reduce the level of surviving lepidopteran pests to very low levels, vastly enhancing the dilution effect offered by the influx of susceptible individuals from refugia, and would thus work in concert to offer a sustainable resistance-management strategy.

It is reasonable to infer from these examples that use of λ-cyhalothrin sprays on Bt Cotton will prolong the life of this valuable technology by slowing the onset of resistance. It is further reasonable to conclude that new technology, be it chemical- or gene-based, will decrease the grower's dependence on established insect-control methods, therefore allowing their use to play to their strengths. Finally, one hypothesis that will be developed further in this paper is that the ideal target for any research programme directed at new insect-control agents will seek effects that are not subject to known mechanisms of resistance.

3. THE INVENTION PROCESS

Invention begins with targets. Companies use targets to focus their invention resources into those areas that will have the most beneficial impact on their business.

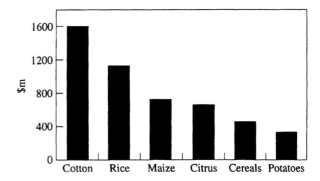

Figure 1. The major insecticide markets, 1996.

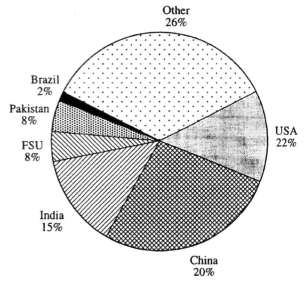

Figure 2. Cotton production, 1996.

Although methods and response to business climates vary from company to company, the targets tend to be similar.

Figure 1 provides some statistics taken from Wood MacKenzie for the largest insecticide markets (Wood MacKenzie 1997). These data considered in isolation would suggest that there is a great deal of value to be found for a new product in a number of markets. However, each of these markets is comprised of a number of pest species over a large range of territories. Furthermore, as alluded to earlier, targets for an individual company will depend on the current products in their range, the fit with products in other market sectors and the company's strategy, among other factors.

To use cotton as an example, although over 60% of pesticide usage is directed at insect control worldwide, the crop is grown on significant acreage in many parts of the world as shown in figure 2. A consequence of this geographical distribution is that the pest spectrum varies considerably from area to area as suggested in table 1.

Although different species, the two most important pests in both the USA and Asia are moths, and the only clear stand-alone target in most areas is for control of heliothine moths. Therefore, when targeting hits on the screens that do not control these key pests, a careful evaluation must be made of the value of the individual

Table 1. *Major cotton pests in the USA and Asia*

USA
  tobacco budworm (*Heliothis virescens*)
  cotton bollworm (*Heliothis zea*)
  boll weevil (*Anthonomus grandis*)
  various bugs (e.g. lygus bugs and stink bugs)

Asia
  cotton bollworm (*Helicoverpa armigera*)
  Earias spp.
  whitefly (e.g. *Bemisia tobaci*)
  jassids, thrips, aphids and mites

segments of the market and the pest spectrum of the particular chemical area at hand. Obviously, combining control of heliothine moths with a larger spectrum of pest control will improve the potential value of the new area of chemistry. Most companies have threshold market values where the minimum size and profitability of the potential market for the new chemical must justify the cost of bringing the chemistry to market.

Figure 3 provides a representation of a generic invention process as might typically be operated in most agrochemical companies. The bars represent increasingly difficult hurdles for an individual project or area of chemistry to surmount. The names provided are a guide to the nature of activities at each stage and example activities are included. This is by no means an exhaustive list of the many scientific inputs required at the various stages and should be considered as indicative only. The process suggested here would cover all of a company's activities, and work directed at insect control might encompass only one-third of the compounds suggested at the later stages. The process is described briefly as follows.

Screening. Most major agrochemical companies screen somewhere between 10 000 and 100 000 compounds a year from various sources. Insecticide screens will typically be run on between three and ten species that are deemed representative of the most important groups of pests selected on the basis of targets. Relatively few compounds show activity on the screens and active compounds are assessed for potency. This screening, combined with assessment of the spectrum of activity, comparison to known classes, possibly the synthesis of a few analogues, and ownership by relevant individuals in the organization, will typically be sufficient to allow progression to the second stage.

Confirmation. Between 10 and 50 analogues will often be prepared during confirmation to determine the chemical scope of the area. To be considered for further progression, an area of chemistry will usually need to demonstrate some breadth of the activity signal (spectrum) and a suggestion that potency can be improved. Alongside this activity will be a thorough glasshouse and laboratory characterization of the better analogues. In addition, preliminary thinking, data collection, and possibly some limited laboratory work around issues relevant to toxicology, ecotoxicology and mode of action will take place. Finally, a preliminary assessment of the potential business case will be made.

The confirmation stage will often last between six months and one year for most areas of chemistry.

Clarification. Relatively few new series pass the next hurdle. At this stage a large, multidisciplinary project will be formed with the hopeful outcome that the research team will be able to identify a single compound for development. Many diverse scientific inputs are required to understand the potential of a new area. For insecticide candidates, an attempt will be made to determine the mode of action and assess the potential for cross-resistance or resistance development.

Development. It will usually require an investment of US $25–100 million to bring a new insecticide to the market. The breadth of the required activities and functional skills is enormous. A few are mentioned in figure 3.

Several key questions relative to resistance need to be posed at the decision points in this process.

1. *Does the prospective area of chemistry rely on a known mode of action?* If so, does the lead have the potential to deliver second-generation performance, as evidenced by better potency, increased spectrum, etc.? If second-generation performance is lacking, it often will not be profitable for an agrochemical company to further develop such a lead, as insufficient market share is likely to be captured to justify the development costs. When second-generation performance is present, factors must be assessed which will differentiate the chemistry from known compounds sharing the mode of action. These include the potential for altered binding at the active site or possible differences in susceptibility to metabolic inactivation mechanisms. The recently introduced sodium-channel binder, indoxacarb, illustrates this point. Although targeted at *Heliothis* control and sharing a general mode of action with the pyrethroids, indoxacarb has been shown to bind the receptor differently than pyrethroids (Wing *et al.* 1998).

2. *Are products that share an established mode of action used in the contemplated market?* Imidacloprid, a nicotinic agonist, brought a novel mode of action to the sucking-pest market when introduced in the early 1990s. Although exhibiting a considerable spectrum of activity, imidacloprid lacked potent activity on certain important classes of insects such as Lepidoptera pests of cotton. Only now are we beginning to see the emergence of the next generation of nitromethylene compounds, the chemical class encompassing imidacloprid, which have a much broader spectrum of activity. Although resistance could develop to imidacloprid in those markets that were established early, it is not unreasonable to expect that market opportunities on new pests will be unaffected by any such resistance.

3. *When the mode of action appears to be novel, is it possible to realistically assess the potential for resistance development?* Limited tools exist to assess some general mechanisms, e.g. whole organisms with elevated ability to metabolize xenobiotics. The ability of these methods to predict the potential impact of metabolic mechanisms on new classes of chemistry requires assumptions based on model systems. Furthermore, few options are currently available to assess the potential for resistance to develop or to predict its extent, evolution and expansion for novel modes of action, even when the site of action is known. However, great promise appears to exist in emerging science.

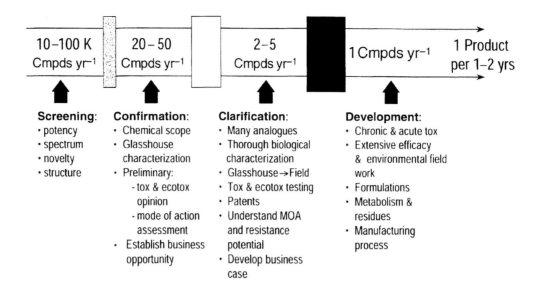

Figure 3. The generic invention process.

Emerging genomic science offers new opportunities to use organisms such as *Drosophila melanogaster* or *Caenorhabditis elegans* to pursue studies of modes of action and to explore the potential for resistance development. The mapping and sequencing of genomes in these species is becoming well advanced. A variety of techniques can be applied to study genetic effects in response to the application of a specific chemical to whole organisms. Panels of organisms that have characterized, genetically based modifications exist in some species (e.g. yeast). Although developments in these areas have been largely driven by medicinal applications, applications to the study of agrochemicals are being increasingly recognized and pursued.

Determining the mode of action of insecticides has important implications for other aspects of agrochemical development. This information can be very useful to secure product registrations and for product stewardship, for example by answering questions such as: is the mode of action unique to insects; and is the mode of toxicity the same as the mode of insecticidal activity?

Although not exhaustive, this treatment should provide the reader with a view of the type of questions that can be posed and studied to facilitate decision-making around invention of new insecticides. The remaining discussion will focus on the two key points raised earlier, which are important in assessing the commercial potential of a new area of insecticidal chemistry: (i) an ability to characterize new chemistry against known modes of action; and (ii) an ability to understand the potential for resistance development with novel or unknown modes of action. These points will be discussed, again referring to the generic invention process represented in figure 3 and by highlighting where an understanding of resistance will come into play.

It is clear that understanding the mode of action of an insecticide can be beneficial very early in invention. When a lead is first screened, it is likely that information that is relevant to the mode of action will be secured from data such as symptoms in affected animals or *in vitro* information collected as a routine part of the screening programme. Most companies have a large data bank of these *in vitro* screens, which are run either as a routine part of the screen or once a hit is identified. These are frequently run on an exclusive basis, that is, to exclude certain known modes of action from further consideration.

Known susceptibility to resistance development can be an important factor in the decision to put a screen in place. For example, for many companies the knowledge that a new area of chemistry had the same mode of action as the cyclodiene insecticides would be a potential fatal negative. Widespread site insensitivity can be found to the cyclodiene chemistry in many species, and is particularly acute in public-health pests. It is therefore likely that cross-resistance would develop in some species to any new area of chemistry that was commercialized with this effect. An example in an area of aryl heterocycles worked on in my company a few years ago (Whittle *et al.* 1995). These compounds, shown in figure 4, were particularly effective on public-health pests. Although taken to a fairly advanced stage, the area was ultimately abandoned, primarily because of growing concerns over resistance to these 'GABAergic' insecticides.

Nonetheless, the recent highly successful introduction of fipronil can be contrasted with this example. Although the threat of cyclodiene cross-resistance is present, clearly the strength of fipronil on non-public-health pests has made this compound a huge commercial success in agricultural applications.

In vitro screens can also be used at an early stage to identify specific desirable modes of action. These screens may be used routinely to complement information generated from whole-organism assays or they may be used to follow up hits from these screens. The philosophy here seems to differ substantially between companies, with some believing that only *in vivo* hits are worth

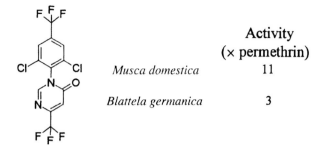

	Activity (× permethrin)
Musca domestica	11
Blattela germanica	3

Figure 4. GABAergic aryl heterocycles.

following up, and others putting substantial investments into identifying and validating novel modes of action, developing *in vitro* screens based on these modes of action and putting large numbers of compounds across these screens.

Many factors will go into defining the value of *in vitro* mode of action screens, but one will be the potential for resistance development. These screens will often be based on a novel mode of action that may or may not be well characterized. For instance, they could be based on the mode of action of an insecticidal natural product or a protein toxin. Regardless, it is important to understand the potential for resistance development to the mode of action. However, this can be an immensely difficult subject, and it is aggravated when dealing with poorly characterized modes of action.

Genomic science seems to be delivering potential tools to meet the challenges here. Characterization of the genomes of relevant species is becoming well advanced and functional genomics are likely to soon provide indicative arrays of all the genes in an insect genome in model systems. For instance, it is possible to imagine arrays on plates where all the genes in *Drosophila* are expressed in *Escherichia coli* and provide a colorimetric response when functioning normally. Application of an insecticidal natural product would stop the colorimetric response of affected genes and possibly associated genes in, for example, a metabolic pathway. Such arrays are currently available for yeast.

In the longer-term, we may be able to use these genetic approaches to group mutants containing specific lesions relevant to important general mechanisms of resistance, such as metabolic mechanisms and reduced uptake or increased excretion. Regarding the ability of an organism to develop site insensitivity, which is arguably the most important resistance mechanism, there seems to be the potential to use genomic science to assess the inclination of the specific target site to adapt to effectors of the active site through a variety of techniques, such as assessing the natural abundance of variation in model systems.

Clearly, many considerations relate to the mode of action of a new area of chemistry and the potential of pests to develop resistance to that mode of action can have an important influence on choices made at the early stages of the invention process. As indicated in figure 3, there are many decisions to be made at these early stages and only a small fraction of potential leads can be developed, given the cost of resources in the subsequent activities.

Resistance will be equally influential on decision-making once new chemistry has progressed through the confirmation/clarification boundary. At this point, one will generally know whether the new insecticidal chemistry has (i) a known mode of action; (ii) an unknown mode of action that is partly characterized (e.g. it is known that active compounds interfere with insect development); or (iii) an unknown and likely novel mode of action that may be difficult to characterize. This information will probably form a critical success factor for the project. A critical success factor can be defined as a factor upon which progression to the next stage of the process is strictly dependent.

When the mode of action is known, a key critical success factor will be to understand the potential for cross-resistance before taking a decision to initiate expensive development activities such as chronic toxicology. Characterizing the potential for cross-resistance can be a challenge. For example, significant resistance to established products that share the mode of action may not have been observed yet in the field. Such resistance may, however, develop during the four to eight years required to bring the new chemistry to the market. In addition, even when resistance has been observed in the field it may be difficult to reproduce measurable effects in the laboratory with field-collected populations, for example because of the influence of environmental factors.

If some evidence of the basis of the mode of action exists, but it is not well understood (e.g. the new chemistry affects a metabolic pathway but the exact site of action is not known), an important critical success factor for the project is likely to be a determination of the exact site of action and understanding of the implications for resistance development.

Surprisingly frequent, however, will be the case where there are few clues regarding the mode of action at this stage. An example of this point is the natural product lepicidin, the basis of the new Lepidoptera-specific insecticide TRACER®. This product appears to have been progressed with no clear understanding of the mode of action and the assumption of novelty. Some information has recently been published in this area (Salgado *et al.* 1997).

In these cases of poorly understood modes of action, a sensible way for projects to proceed centres on collecting empirically derived data in key areas that will raise concerns during development and registration. These include acute testing on birds and *Daphnia*, indicative toxicology screens to supplement the typical acute toxicology required to distribute samples, and an assessment of soil mobility and persistence to determine if any groundwater issues are likely to arise. Such studies will highlight areas of concern. In the meantime, the project will be able to continue to pursue reasonable studies with the objectives of determining the mode of action and the potential for resistance development.

4. CONCLUSIONS

This discussion has focused on what is carried out in the agrochemical industry to use information on mode of action and resistance to facilitate the process of inventing new insecticides. However, most of the science practised involves the application of basic research from universities and private research institutions. Companies heavily rely on these fundamental studies as key sources of essential knowledge in areas such as mode of action, site of action, receptors, enzymes, metabolism, rational design, new receptor screens and genetics at a molecular level, just to mention a few areas that are important to the invention process. This fundamental work has immense importance in ensuring our continued ability to secure the new products required by agriculture.

It is also important, at this point, to put the main arguments made in this paper in context. Readers should understand that the potential for cross-resistance and the desire to introduce new modes of action do not dominate an industry's thinking in assessing the commercial potential of a new insecticide candidate. In fact, other factors—does it work? will it make money? is it safe?—will have an overriding importance and will encompass an evaluation of the possibility of cross-resistance. It is also important to remember that no product has totally lost its usefulness solely as a result of resistance development.

Nevertheless, the future will undoubtedly see a better balance of transgenic, biological and chemical means of insect control which reduces the selection pressure against any one product. Although industry has taken substantial steps towards managing resistance development through techniques such as reserving the use of the best broad-spectrum products for the periods of peak pest infestation, a superior understanding of the nature of resistance will certainly allow science to develop in managing resistance. Agriculture is heading toward a future state of integrated crop management where the best available products based on chemical, biotechnological and biological control are selectively and precisely brought to bear through superior monitoring and information techniques to fight infestations in an optimal way. In this regard, the earlier points regarding Bollgard[®] cotton are important. Not only will this Bt-based transgenic technology prove significant in prolonging the life of established chemical products, but the judicious use of chemicals in combination with Bt crops are likely to extend the effectiveness of this new technology.

New insect-control technology is difficult to find and costly to bring to market. When introduced, it is welcomed and quickly embraced by growers in numerous market segments. In many markets, a primary driving force is the belief that development of resistance is typically inevitable. Therefore, the fundamental studies undertaken at universities and research institutions, which underpin the efforts of the agrochemical industry, are critical in the invention and development of new insecticides.

REFERENCES

Graves, J. B., Leonard, B. R., Micinski, S., Burris, E. & Martin, S. H. 1993 Monitoring insecticide resistance in tobacco budworm and bollworm in Louisiana. In *Proceedings of the Beltwide Cotton Production Conference*, pp. 788–790. Memphis, TN: National Cotton Council.

Harris, J. G. 1998 The usage of Karate[®] (A-cyhalothrin) over-sprays in combination with refugia, as a viable and sustainable resistance management strategy for Bt cotton. In *Proceedings of the Beltwide Cotton Production Conference*, pp. 1217–1221. Memphis, TN: National Cotton Council.

Mink, J., Harrison & Martin, S. 1997 Performance and benefits of Karate[®] insecticide on Bollgard[®] cotton. In *Proceedings of the Beltwide Cotton Production Conference*, pp. 898–899. Memphis, TN: National Cotton Council.

Oerke, E. C., Dehne, H. W., Schoenbeck, F. & Weber, A. 1994 *Crop production and crop protection: estimated losses in major food and cash crops.* Amsterdam and New York: Elsevier.

Salgado, V. L., Watson, G. B. & Sheets, J. J. 1997 Studies on the mode of action of spinosad, the active ingredient in tracer insect control. In *Proceedings of the Beltwide Cotton Production Conference*, pp. 1082–1084. Memphis, TN: National Cotton Council.

Whittle, A. J., Fitzjohn, S., Mullier, G., Pearson, D. P. J., Perrior, T. R., Taylor, R. & Salmon, R. 1995 The use of computer-generated electrostatic surface maps for the design of new 'GABA-ergic' insecticide. *Pestic. Sci.* **44**, 29–31.

Wing, K. D., Schnee, M. E., Sacher, M. & Connair, M. 1998 A novel oxadiazine insecticide is bioactivated in Lepidopteran larvae. *Arch. Insect Biochem. Physiol.* **37**, 91–103.

Wood MacKenzie 1977 Update of products section. Reference volume of the agrochemical service, pp. 36–49.

Predicting insecticide resistance: mutagenesis, selection and response

J. A. McKenzie[*] and P. Batterham

Department of Genetics, University of Melbourne, Parkville, Victoria 3052, Australia

Strategies to manage resistance to a particular insecticide have usually been devised after resistance has evolved. If it were possible to predict likely resistance mechanisms to novel insecticides before they evolved in the field, it might be feasible to have programmes that manage susceptibility. With this approach in mind, single-gene variants of the Australian sheep blowfly, *Lucilia cuprina*, resistant to dieldrin, diazinon and malathion, were selected in the laboratory after mutagenesis of susceptible strains. The genetic and molecular bases of resistance in these variants were identical to those that had previously evolved in natural populations. Given this predictive capacity for known resistances, the approach was extended to anticipate possible mechanisms of resistance to cyromazine, an insecticide to which *L. cuprina* populations remain susceptible after almost 20 years of exposure. Analysis of the laboratory-generated resistant variants provides an explanation for this observation. The variants show low levels of resistance and a selective advantage over susceptibles for only a limited concentration range. These results are discussed in the context of the choice of insecticides for control purposes and of delivery strategies to minimize the evolution of resistance.

Keywords: genetic response; insecticide resistance; *Lucilia cuprina*; mutagenesis; selection

1. INTRODUCTION

Effective pesticides play a key role in the management of agricultural ecosystems. For example, in sheep the control of ectoparasites is currently dependent on the use of chemicals. This situation will continue to apply in, at least, the intermediate term (Levot 1993). Pesticide usage leads to chemical residues, resulting in issues of occupational health and safety and, particularly after processing of the fleece, potential environmental degradation. Similar difficulties arise in other agricultural systems.

The residue problem is exacerbated by the evolution of resistance to the pesticide. To maintain control, the response is commonly more frequent application of higher concentrations of pesticide (Daly & McKenzie 1986; Roush & Tabashnik 1990; Denholm & Rowland 1992; McKenzie 1996) with resistance-management strategies usually based on general models influenced by the anticipated genetic basis of resistance (Georghiou & Taylor 1977a,b; Roush 1989; Tabashnik 1990; Rosenheim & Tabashnik 1990; Gressel 1995). Specific strategies are generally put in place only after resistance has already evolved (Daly & McKenzie 1986; Forrester *et al.* 1993).

There would be many advantages if a likely resistance mechanism could be predicted before a new insecticide was introduced into the field. The availability of genetic, toxicological, biochemical, cross-resistance and relative fitness data would maximize the chance of devising strategies to minimize the evolution of resistance, that is, enable the management of susceptibility (Leeper *et al.*

1986; Daly & McKenzie 1986; Firko & Hayes 1990; Tabashnik 1990; McKenzie 1996). To attempt to make a meaningful prediction, it is essential to have an understanding of the genetic basis underlying the observed phenotypic basis of resistance. It is on this variation that evolutionary processes will act (Roush & McKenzie 1987).

2. THE GENETIC BASIS OF RESISTANCE

Monogenic and polygenic control of resistance is observed in natural populations (Roush & McKenzie 1987; McKenzie 1996). There is considerable debate about the relative importance of these mechanisms (Roush & McKenzie 1987; Mallet 1989; Macnair 1991; McKenzie & Batterham 1994, 1995; Groeters 1995; Tabashnik 1995; McKenzie 1996). Such discussions are a subset of a more general evolutionary debate concerning monogenic and polygenic responses during adaptation (Lande 1983; Macnair 1991; Orr & Coyne 1992).

For the purpose of this paper, let us assume that when an insecticide is first introduced for pest control the population consists of susceptible phenotypes. Within the susceptibles we also assume that viability is normally distributed with differences between individuals of that continuous distribution being under polygenic control (McKenzie & Batterham 1994). If selection acts within the distribution, the differences between selectively advantaged and disadvantaged individuals will be polygenically based and therefore a polygenic response will be expected. If a rare resistant mutation, outside the range of the original phenotypic distribution, confers a selective advantage over susceptibles, at concentrations that cannot

[*]Author for correspondence (j.mckenzie@genetics.unimelb.edu.au).

be accommodated by a polygenic response, a monogenic response is expected (Whitten & McKenzie 1982; Macnair 1991; McKenzie & Batterham 1994; McKenzie 1996). Comparative laboratory studies in the Australian sheep blowfly, *Lucilia cuprina*, support these propositions (McKenzie *et al.* 1980, 1992). In natural populations of *L. cuprina*, resistances to dieldrin, diazinon and malathion are primarily due to allelic substitutions at single genetic loci, an observation consistent with the way in which natural selection for resistance acts in this species (McKenzie 1993, 1996).

In the light of the above arguments and observations we have attempted to mimic, and predict, field resistance by selecting for monogenic responses in the laboratory.

3. PREDICTING RESISTANCE MECHANISMS

To have any confidence that variants, selected in the laboratory for resistance to a novel insecticide, will be predictive of mechanisms that may evolve in the field after the insecticide is introduced, it is first necessary to conduct the appropriate control experiments. That is, it is necessary to demonstrate that variants selected in the laboratory for resistance to previously used insecticides are equivalent to the resistant variants that evolved in natural populations. This has been done in *L. cuprina*.

(a) *Control experiments: resistance to dieldrin, diazinon and malathion*

Dieldrin, a cyclodiene, was used to control sheep blowfly between 1955 and 1958 (Hughes & McKenzie 1987). Resistance to this insecticide evolved in natural populations of *L. cuprina* within two years as a result of allelic substitution at the *Rdl* locus on chromosome V (McKenzie 1987). Resistance is due to less effective blocking of insect neuronal GABA receptors by the insecticide in resistant strains (Smyth *et al.* 1992). The resistance gene is the orthologue of the dieldrin resistance gene, *Rdl*, of *Drosophila melanogaster*. The same serine-to-alanine substitution is responsible for resistance in both species (P. Batterham, Z. Chen, R. H. ffrench-Constant, K. Freebairn, R. D. Newcomb and J. A. McKenzie, unpublished data; ffrench-Constant 1994).

The organophosphorus insecticide diazinon replaced dieldrin for the control of blowfly in 1958. Resistance, essentially controlled by allelic substitution of the *Rop-1* locus on chromosome IV (McKenzie 1993), was observed in the field in 1965 (Hughes & McKenzie 1987). The susceptible allele of the *Rop-1* gene encodes a carboxylesterase, E3 (Hughes & Devonshire 1982). The resistance substitution (glycine[137] to aspartic acid) leads to a gain of organophosphorus hydrolase activity at the expense of the carboxylesterase activity (Newcomb *et al.* 1997; Campbell *et al.* 1998a). A second gene, *Rop-2*, on chromosome VI, also influences resistance to diazinon, but occurs very rarely in natural populations (Arnold & Whitten 1976). *Rop-2* codes for a mixed-function oxidase (Hughes & Devonshire 1982; Terras *et al.* 1983).

Malathion, also an organophosphorus insecticide, was not used specifically for blowfly control. However, resistance, of *L. cuprina* to this chemical was detected in 1968 and mapped to a gene on chromosome IV (Hughes *et al.* 1984). This gene, *Rmal*, was believed to be tightly linked

Table 1. *The genetic and molecular bases of resistance to dieldrin, diazinon and malathion in laboratory- and field-selected variants of* L. cuprina

(Derived from: Hughes & McKenzie 1987; McKenzie *et al.* 1992; Smyth *et al.* 1992; Newcomb *et al.* 1997; Campbell *et al.* 1998a; P. Batterham, Z. Chen, R. H. ffrench-Constant, K. Freebairn, R. D. Newcomb and J. A. McKenzie, unpublished data.)

resistance	genetic locus		amino-acid change susceptible⟶resistant	
	field	laboratory	field	laboratory
dieldrin	*Rdl*	*Rdl*	ser[302]⟶ala	ser[302]⟶ala
diazinon	*Rop-1*	*Rop-1*	gly[137]⟶asp	gly[137]⟶asp
malathion	*Rmal*	*Rmal*	trp[251]⟶leu	trp[251]⟶leu

to the *Rop-1* locus (Raftos & Hughes 1986; Smyth *et al.* 1994), but recent molecular data suggest that malathion resistance is, in fact, controlled by an allele of the *Rop-1* locus (Campbell *et al.* 1998a). For convenience the *Rmal* notation is maintained. As a result of a single amino-acid substitution (tryptophan[251] to leucine), the resistant form of the E3 enzyme has both organophosphorus hydrolase and malathion carboxylesterase activities (Hughes *et al.* 1984; Campbell *et al.* 1998a,b). Hence, *Rmal* genotypes provide some resistance to diazinon but *Rop-1* genotypes do not provide no resistance to malathion (Campbell *et al.* 1998b).

In the laboratory, susceptible strains were mutagenized by using ethyl methanesulphonate (EMS) and progeny selected for resistance to dieldrin or diazinon by screening above the lethal concentration (LC$_{100}$) of susceptibles. The dieldrin- (Smyth *et al.* 1992) and diazinon- (McKenzie *et al.* 1992) resistant strains generated showed similar toxicological, biochemical, genetic and molecular bases similar to those observed in natural populations (table 1). It can therefore be concluded that if these experiments had been conducted before the insecticides were introduced for blowfly control, the laboratory results would have been highly predictive of the mechanisms that evolved in natural populations. Two comments should be made to place this statement in context.

First, selection in the laboratory for resistance to diazinon generated variants which are resistant to just this chemical (*Rop-1*) and variants resistant to malathion, with some resistance to diazinon (*Rmal*). Second, *Rop-2* variants have not been generated, thus far, in the laboratory mutagenesis–selection regime.

The first observation is consistent with resistance to malathion having been selected in the field by exposure to diazinon. Although other explanations are possible (Campbell *et al.* 1998b), the laboratory result supports the initial hypothesis of Hughes *et al.* (1984) of such a selective process.

The absence of *Rop-2* variants from the laboratory screen may help provide an explanation for its rarity in field populations; however, a larger sample of resistant mutants is necessary for a definitive comment. This work is in progress.

In summary, taking these comments into account, the control experiments provide sufficient encouragement to

Table 2. *Comparison of strains of* L. cuprina *resistant or susceptible to cyromazine*

(Derived from Yen *et al.* (1996).)

strain	resistance ratio[*]	viability	concentration range (% (W/V) × 10^{-5}) of selective advantage[**]
susceptible	1	viable	0–4
Cyr 4 (1)	4.1	recessive lethal	6–30
Cyr 4 (2)	1.5	viable	6–20
Cyr 5 (1)	2.8	recessive subvital	4–25
Cyr 5 (2)	2.2	recessive subvital	8–25

[*]LC_{90} of heterozygote/LC_{90} of susceptible.
[**]Susceptible compared with resistant heterozygote.

extend the laboratory programme to the prediction of possible resistance mechanisms before they have evolved to a particular insecticide in natural populations. An attempt to select for resistance to cyromazine was therefore instigated.

(b) *Predicting resistance to cyromazine*

Cyromazine acts as an insect growth regulator (Binnington 1985) and was introduced during the late 1970s to control a number of pests (Hart *et al.* 1982; Iseki & Georghiou 1986; Hughes *et al.* 1989; Sirota & Grafius 1994). Moderate levels of resistance have been observed in natural populations of *Musca domestica* (Bloomcamp *et al.* 1987; Shen & Plapp 1990; Keiding *et al.* 1992), but resistance is yet to evolve in the field in *L. cuprina* (Levot 1993; Yen *et al.* 1996).

The laboratory mutagenesis and selection regime described above generated four resistant strains (Yen *et al.* 1996). Resistance in each strain was controlled by a single gene. Loci (Cyr4 and Cyr5) mapped to chromosomes IV and V, respectively. Two alleles ((1) and (2)) have been identified at each locus. The resistance-ratio, genetic and viability characterizations of these strains (table 2) help to explain why cyromazine-resistance has been slow to evolve in natural populations of *L. cuprina*. If the laboratory variants are typical of those that arise in the field, the evolutionary window to select for resistance is only just open (Yen *et al.* 1996).

It should be noted that a number of other genes may control resistance to cyromazine. For example, in *Drosophila melanogaster*, Adcock *et al.* (1993) have selected genes different from, and similar to, the ones in *L. cuprina*. The conservation of genetic maps between these species (Weller & Foster 1993) allows this conclusion. More novel variants are expected as the mutagenesis and selection programmes have yet to achieve saturation in either species. However, thus far, all field- and laboratory-selected cyromazine-resistant variants appear to share one characteristic, low resistance ratios (Bloomcamp *et al.* 1987; Shen & Plapp 1990; Keiding *et al.* 1992; Adcock *et al.* 1993; Yen *et al.* 1996). Therefore, there is cause for optimism in the predictive nature of the laboratory experiments in *L. cuprina*. The relatively low LC_{100} of heterozygotes (table 2) and the toxicological properties of cyromazine also allow the possibility of delivery systems to minimize the probability of resistance evolving.

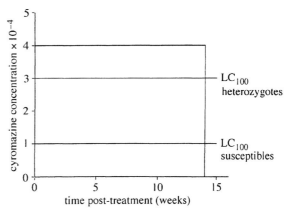

Figure 1. Idealized square-wave decay curve concentration to minimize the evolution of resistance to cyromazine in natural populations of *L. cuprina*. The concentrations for LC_{100} of susceptibles and heterozygotes are derived from Yen *et al.* (1996).

4. INSECTICIDE DELIVERY SYSTEMS

Selection for resistance is minimized when the concentration range over which resistant phenotypes are advantaged, relative to susceptibles, is restricted. When resistance at a single locus first evolves in natural populations, selection is for heterozygotes (Whitten & McKenzie 1982). Therefore, if an insecticide can be delivered at a concentration above the LC_{100} of heterozygotes, with a square-wave decay curve, the probability of resistance evolving can be theoretically reduced to zero (McKenzie 1987, 1996).

Cyromazine is an extremely safe compound to mammals and is taken into the serum after oral delivery in sheep. Through the use of intraruminal capsules, a square-wave decay curve can be generated (Anderson *et al.* 1989). The results from the laboratory variants allow the LC_{100} of susceptibles and heterozygotes to be identified. If these data were available before the release of the chemical for protection from blowfly, use of the intraruminal system, at the appropriate concentration, would allow control with little chance of resistance evolving (figure 1).

5. DISCUSSION

The laboratory experiments have demonstrated that it is possible to select for single-gene resistance by screening above the LC_{100} of susceptibles. Mutagens have been commonly used to enhance mutation rates but their use in insecticide-resistance studies has been restricted since the initial studies of Kikkawa (1964; McKenzie 1996).

The spontaneous rate of mutation from susceptible to resistant alleles is not known. Theoretical arguments can be made for rates in the range 10^{-3} to 10^{-13} (Whitten & McKenzie 1982). In the mutagenesis and selection experiments described, the average rate, after mutagenesis, was of the order of 4×10^{-6}. No resistant mutants arose spontaneously in unmutagenized cultures (McKenzie *et al.* 1992; Smyth *et al.* 1992; Yen *et al.* 1996) and therefore, EMS mutagenesis has increased the variation available to be screened by selection.

In the control experiments, the resistance to dieldrin, diazinon and malathion generated in the laboratory was identical to that that evolved in the field (table 1). This is perhaps not surprising in the case of resistance to dieldrin, as the molecular basis of resistance has been due to parallel mutations over a number of species (ffrench-Constant 1994). However, for resistance to the organophosphorus insecticides mechanisms involving acetylcholinesterases, carboxylesterases, mixed-function oxidases and glutathione-S-transferases have been recorded in the literature (Russell *et al.* 1990; McKenzie 1996). Resistances based on carboxylesterases and, rarely, on mixed-function oxidases have been recorded in natural populations of *L. cuprina* (Hughes & Devonshire 1982; Terras *et al.* 1983) but, only the former have been generated in the laboratory (McKenzie *et al.* 1992). It remains to be seen whether further laboratory selection results in other variants but, in the context of this paper, the control experiments have been successful. They provide a foundation for the use of the mutagenesis and selection approach to predict, and allow analysis of, resistance mechanisms to novel insecticides before the chemicals are introduced into the field.

The results for the cyromazine-resistant variants (table 2) are also encouraging in this regard. The low-resistance ratios generated in the laboratory strains of *L. cuprina* (Yen *et al.* 1996) and *D. melanogaster* (Adcock *et al.* 1993) match those observed in natural populations of *M. domestica* (Bloomcamp *et al.* 1987; Shen & Plapp 1990; Keiding *et al.* 1992). The availability of resistant variants also allows the current lack of knowledge about the actual physiological and molecular foundations of resistance to cyromazine (Binnington 1985; Friedel & McDonell 1985; Kotze & Reynolds 1990) to be addressed. Molecular and genetic conservation across species provide considerable flexibility of the analysis of resistance systems (Oakeshott & Whitten 1992; ffrench-Constant 1994; Severson *et al.* 1997). Therefore, model systems may have an important role in predicting resistance mechanisms. *D. melanogaster* is already commonly used in resistance studies (McKenzie 1996) and there are compelling genetic and molecular arguments for its use as a model organism (Wilson 1988). This species has already proved useful in studies of laboratory mutagenesis and selection for resistance studies (Kikkawa 1964; Wilson & Fabian 1986; Adcock *et al.* 1993). Its role is likely to expand.

Information about the LC_{100} of resistant heterozygotes allows key delivery concentrations to be defined (figure 1) before a chemical is released. Relative fitness estimates can be made before resistance evolves in the field (Yen *et al.* 1996) and the selective concentration range determined (table 2). Such data have helped explain the relative rates of development of resistance to different insecticides (Roush & McKenzie 1987; Roush & Tabashnik 1990; McKenzie 1996) after resistance has evolved in the field. They will be equally important in determining the application strategies for a new insecticide if possible resistant mechanisms are predicted.

By way of example, if the laboratory experiments for dieldrin, diazinon and cyromazine had been performed before the chemicals were used for blowfly control, it would have been predicted that dieldrin-resistance would develop most rapidly. Furthermore, the level of resistance would render the insecticide ineffective for blowfly control.

Diazinon resistance would have been predicted to evolve more slowly and it would have been predicted that, even in the face of high *Rop-1* allelic frequencies, the insecticide would provide adequate protection against blowfly until a replacement insecticide was available. Cyromazine resistance would have been predicted to have relatively little chance of evolving. So far each of these predictions would have proved to be accurate (McKenzie 1996).

The strategy of selecting for resistance in the laboratory before a chemical is released, therefore, appears promising if our aim is to anticipate likely mechanisms before they evolve in natural populations. This information is important if we are to maximize the probability of managing susceptibility, an option preferable option to attempting to manage resistance (Daly & McKenzie 1986).

Janet Yen, Phillip Daborn and Kris Freebairn are thanked for helpful discussions. The Australian Research Council and the International Wool Secretariat provided financial support.

REFERENCES

Adcock, G. J., Batterham, P., Kelly, L. E. & McKenzie, J. A. 1993 Cyromazine resistance in *Drosophila melanogaster* (Diptera: Drosophilidae) generated by ethyl methanesulfonate mutagenesis. *J. Econ. Entomol.* **86**, 1001–1008.

Anderson, N., McKenzie, J. A., Laby, R. H., Strong, M. B. & Jarrett, R. G. 1989 Intraruminal controlled release of cyromazine for the prevention of *Lucilia cuprina* myiasis in sheep. *Res. Vet. Sci.* **46**, 131–138.

Arnold, J. T. A. & Whitten, M. J. 1976 The genetic basis for organophosphorus resistance in the Australian sheep blowfly, *Lucilia cuprina* (Weidemann) (Diptera: Calliphoridae). *Bull. Entomol. Res.* **66**, 561–568.

Binnington, K. C. 1985 Ultrastructural changes in the cuticle of the sheep blowfly, *Lucilia cuprina*, induced by certain insecticides and biological inhibitors. *Tissue & Cell* **17**, 131–140.

Bloomcamp, C. L., Patterson, R. S. & Koehler, P. G. 1987 Cyromazine resistance in the house fly (Diptera: Muscidae). *J. Econ. Entomol.* **80**, 352–357.

Campbell, P. M., Newcomb, R. D., Russell, R. J. & Oakeshott, J. G. 1998*a* Two different amino acid substitutions in the ali-esterase, E_3, confer alternative types of organophosphorus insecticide resistance in the sheep blowfly, *Lucilia cuprina*. *Insect Biochem. Molec. Biol.* **28**, 139–150.

Campbell, P. M., Yen, J. L., Masoumi, A., Russell, R. J., Batterham, P., McKenzie, J. A. & Oakeshott, J. G. 1998*b* Cross-resistance patterns among *Lucilia cuprina* (Diptera: Calliphoridae) resistant to organophosphorus insecticides. *J. Econ. Entomol.* **91**, 367–375.

Daly, J. C. & McKenzie, J. A. 1986 Resistance management strategies in Australia. The *Heliothis* and 'Wormkill' programmes. In *Proceedings 1986 Brighton Crop Protection Conference*, pp. 951–959. Croydon, UK: British Crop Protection Council.

Denholm, I. & Rowland, M. W. 1992 Tactics for managing pesticide resistance in arthropods. *A. Rev. Entomol.* **37**, 91–112.

ffrench-Constant, R. H. 1994 The molecular and population genetics of cyclodiene insecticide resistance. *Insect Biochem. Molec. Biol.* **24**, 335–345.

Firko, M. J. & Hayes, J. L. 1990 Quantitative genetic tools for insecticide risk assessment: estimating the heritability of resistance. *J. Econ. Entomol.* **83**, 647–654.

Forrester, N. W., Cahill, M., Bird, L. J. & Layland, J. 1993 Management of pyrethroid and endosulphan resistance in *Helicoverpa armigera* (Lepidoptera: Noctuidae) in Australia. *Bull. Entomol. Res.* (Suppl.) **1**, 1–132.

Friedel, T. & McDonell, P. A. 1985 Cyromazine inhibits reproduction and larval development of the Australian sheep blowfly. *J. Econ. Entomol.* **78**, 868–873.

Georghiou, G. P. & Taylor, C. E. 1977a Genetic and biological influences in the evolution of insecticide resistance. *J. Econ. Entomol.* **70**, 319–323.

Georghiou, G. P. & Taylor, C. E. 1977b Operational influences in the evolution of insecticide resistance. *J. Econ. Entomol.* **70**, 653–658.

Gressel, J. 1995 Catch 22—mutually exclusive strategies for delaying/preventing polygenically vs monogenically inherited resistances. In *Options 2000* (ed. N. Ragsdale), pp. 1–10. Washington, DC: ACS.

Groeters, F. R. 1995 Insecticide resistance. *Trends Ecol. Evol.* **10**, 164.

Hart, R. J., Cavey, W. A., Ryan, K. J., Strong, M. B., Moore, B., Thomas, P. L., Boray, J. C. & Von Orelli, M. 1982 CGA-72662, a new sheep blowfly insecticide. *Aust. Vet. J.* **59**, 104–109.

Hughes, P. B. & Devonshire, A. L. 1982 The biochemical basis of resistance to organophosphorus insecticides in the sheep blowfly, *Lucilia cuprina*. *Pestic. Biochem. Physiol.* **18**, 289–297.

Hughes, P. B. & McKenzie, J. A. 1987 Insecticide resistance in the Australian sheep blowfly, *Lucilia cuprina*: speculation, science and strategies. In *Combating resistance to xenobiotics. Biological and chemical approaches* (ed. M. G. Ford, D. W. Holloman, B. P. S. Khambay & R. M. Sawicki), pp. 162–177. Chichester, UK: Ellis Horwood.

Hughes, P. B., Green, P. E. & Reichmann, K. G. 1984 Specific resistance to malathion in laboratory and field populations of the Australian sheep blowfly, *Lucilia cuprina* (Diptera: Calliphoridae). *J. Econ. Entomol.* **77**, 1400–1404.

Hughes, P. B., Dauterman, W. C. & Motoyama, N. 1989 Inhibition of growth and development of tobacco hornworm (Lepidoptera: Sphiangidae) larvae by cyromazine. *J. Econ. Entomol.* **82**, 45–51.

Iseki, A. & Georghiou, G. P. 1986 Toxicity of cyromazine to strains of the housefly (Diptera: Muscidae) variously resistant to insecticides. *J. Econ. Entomol.* **79**, 1192–1195.

Keiding, J., El-Khodary, A. S. & Jespersen, J. B. 1992 Resistance risk assessment of two insect development inhibitors, diflubenzuron and cyromazine, for control of the house fly *Musca domestica* L. II. Effect of selection pressure in laboratory and field populations. *Pestic. Sci.* **35**, 27–37.

Kikkawa, H. 1964 Genetical studies on the resistance to parathion in *Drosophila melanogaster*. II. Induction of a resistance gene from its susceptible allele. *Botyu-Kagaku* **29**, 37–42.

Kotze, A. C. & Reynolds, S. E. 1990 Mechanical properties of the cuticle of *Manduca sexta* larvae treated with cyromazine. *Pestic. Biochem. Physiol.* **38**, 267–272.

Lande, R. 1983 The response to selection on major and minor mutations affecting a metrical trait. *Heredity* **50**, 47–65.

Leeper, J. R., Roush, R. T. & Reynolds, H. T. 1986 Preventing or managing resistance in arthropods. In *Pesticide resistance: strategies and tactics for management*, pp. 335–346. Washington: National Academy Press.

Levot, G. W. 1993 Insecticide resistance: new developments and future options for fly and lice control on sheep. *Wool Technol. Sheep Breed.* **41**, 108–119.

McKenzie, J. A. 1987 Insecticide resistance in the Australian sheep blowfly—messages for pesticide usage. *Chem. Ind.* **8**, 266–269.

McKenzie, J. A. 1993 Measuring fitness and intergenic interactions: the evolution of resistance to diazinon in *Lucilia cuprina*. *Genetica* **90**, 227–237.

McKenzie, J. A. 1996 *Ecological and evolutionary aspects of insecticide resistance*. Austin, TX: R. G. Landes–Academic Press.

McKenzie, J. A. & Batterham, P. 1994 The genetics, molecular and phenotypic consequences of selection for insecticide resistance. *Trends Ecol. Evol.* **9**, 166–169.

McKenzie, J. A. & Batterham, P. 1995 Insecticide resistance—reply. *Trends Ecol. Evol.* **10**, 165.

McKenzie, J. A., Dearn, J. M. & Whitten, M. J. 1980 Genetic basis of resistance to diazinon in Victorian populations of the Australian sheep blowfly, *Lucilia cuprina*. *Aust. J. Biol. Sci.* **33**, 85–95.

McKenzie, J. A., Parker, A. G. & Yen, J. L. 1992 Polygenic and single gene responses to selection for resistance to diazinon in *Lucilia cuprina*. *Genetics* **130**, 613–620.

Macnair, M. R. 1991 Why the evolution of resistance to anthropogenic toxins normally involves major gene changes: the limits to natural selection. *Genetica* **84**, 213–219.

Mallet, J. 1989 The evolution of insecticide resistance: have the insects won? *Trends Ecol. Evol.* **4**, 336–340.

Newcomb, R. D., Campbell, P. M., Ollis, D. L., Cheah, E., Russell, R. J. & Oakeshott, J. G. 1997 A single amino acid substitution converts a carboxylesterase organophosphorus hydrolase and confers insecticide resistance on a blowfly. *Proc. Natn. Acad. Sci. USA* **94**, 7464–7468.

Oakeshott, J. G. & Whitten, M. J. (ed.) 1992 *Molecular approaches to fundamental and applied entomology*. New York: Springer.

Orr, H. A. & Coyne, J. A. 1992 The genetics of adaptation: a reassessment. *Am. Nat.* **140**, 725–742.

Raftos, D. A. & Hughes, P. B. 1986 Genetic basis of specific resistance to malathion in the Australian sheep blowfly, *Lucilia cuprina* (Diptera: Calliphoridae). *J. Econ. Entomol.* **79**, 553–557.

Rosenheim, J. A. & Tabashnik, B. E. 1990 Evolution of pesticide resistance: interactions between generation time and genetic, ecological and operational factors. *J. Econ. Entomol.* **83**, 1184–1193.

Roush, R. T. 1989 Designing resistance management programs: how can you choose? *Pestic. Sci.* **26**, 423–441.

Roush, R. T. & McKenzie, J. A. 1987 Ecological genetics of insecticide and acaricide resistance. *A. Rev. Entomol.* **32**, 361–380.

Roush, R. T. & Tabashnik, B. E. (ed.) 1990 *Pesticide resistance in arthropods*. New York: Chapman & Hall.

Russell, R. J., Dumanic, M. C., Foster, G. G., Weller, G. L., Healy, M. J. & Oakeshott, J. G. 1990 Insecticide resistance as a model system for studying molecular evolution. In *Ecological and evolutionary genetics of Drosophila* (ed. J. S. F. Barker, W. T. Starmer & R. J. MacIntyre), pp. 293–314. New York: Plenum.

Severson, D. W., Anthony, N. M., Andreev, O. & ffrench-Constant, R. H. 1997 Molecular mapping of insecticide resistance genes in the yellow fever mosquito (*Aedes aegypti*). *J. Hered.* **88**, 520–524.

Shen, J. & Plapp, F. W. 1990 Cyromazine resistance in the house fly: genetics and cross resistance to diflubenzuron. *J. Econ. Entomol.* **83**, 1689–1697.

Sirota, J. M. & Grafius, E. 1994 Effects of cyromazine on larval survival, pupation, and adult emergence of Colorado potato beetle (Coleoptera: Chrysomelidae). *J. Econ. Entomol.* **87**, 577–582.

Smyth, K.-A., Parker, A. G., Yen, J. L. & McKenzie, J. A. 1992 Selection of dieldrin-resistant strains of *Lucilia cuprina* (Diptera: Calliphoridae) after ethyl methanesulphonate mutagenesis of a susceptible strain. *J. Econ. Entomol.* **85**, 352–358.

Smyth, K.-A., Russell, R. J. & Oakeshott, J. G. 1994 A cluster of at least three esterase genes in *Lucilia cuprina* includes malathion carboxylesterase and two other esterases implicated in resistance to organophosphates. *Biochem. Genet.* **32**, 437–453.

Tabashnik, B. E. 1990 Modeling and evaluation of resistance management tactics. In *Pesticide resistance in arthropods* (ed. R. T. Roush & B. E. Tabashnik), pp. 153–182. New York: Chapman & Hall.

Tabashnik, B. E. 1995 Insecticide resistance. *Trends Ecol. Evol.* **10**, 164–165.

Terras, M. A., Rose, H. A. & Hughes, P. B. 1983 Aldrin epoxidase activity in larvae of a susceptible and a resistant strain of the sheep blowfly, *Lucilia cuprina* (Wied.). *J. Aust. Entomol. Soc.* **22**, 256.

Weller, G. L. & Foster, G. G. 1993 Genetic maps of the sheep blowfly *Lucilia cuprina*: linkage group correlations with other dipteran genera. *Genome* **36**, 495–506.

Whitten, M. J. & McKenzie, J. A. 1982 The genetic basis for pesticide resistance. In *Proceedings of the 3rd Australasian Conference of Grassland Invertebrates* (ed. K. E. Lee), pp. 1–16. Adelaide: South Australian Government Printer.

Wilson, T. G. 1988 *Drosophila melanogaster* (Diptera: Drosophilidae): a model insect for insecticide resistance studies. *J. Econ. Entomol.* **88**, 22–27.

Wilson, T. G. & Fabian, J. A. 1986 *Drosophila melanogaster* mutant resistant to a chemical analog of juvenile hormone. *Devl Biol.* **118**, 190–201.

Yen, J. L., Batterham, P., Gelder, B. & McKenzie, J. A. 1996 Predicting resistance and managing susceptibility to cyromazine in the Australian sheep blowfly *Lucilia cuprina*. *Aust. J. Exp. Agric.* **36**, 413–420.

Resistance to insecticides in heliothine Lepidoptera: a global view

Alan R. McCaffery

Zoology Division, School of Animal and Microbial Sciences, and Crop Protection Unit, The University of Reading, Whiteknights, PO Box 228, Reading RG6 6AJ, UK (a.r.mccaffery@reading.ac.uk)

The status of resistance to organophosphate, carbamate, cyclodiene and pyrethroid insecticides in the heliothine Lepidoptera is reviewed. In particular, resistance in the tobacco budworm, *Heliothis virescens*, and the corn earworm, *Helicoverpa zea*, from the New World, and the cotton bollworm, *Helicoverpa armigera*, from the Old World, are considered in detail. Particular emphasis has been placed on resistance to the most widely used of these insecticide groups, the pyrethroids. In each case, the incidence and current status of resistance are considered before a detailed view of the mechanisms of resistance is given. Controversial issues regarding the nature of mechanisms of resistance to pyrethroid insecticides are discussed. The implications for resistance management are considered.

Keywords: insecticide resistance; Heliothinae; *Heliothis virescens*; *Helicoverpa armigera*; mechanisms of resistance

1. INTRODUCTION

Lepidopteran species in the genera *Heliothis* and *Helicoverpa* are grouped together in the Trifine subfamily Heliothinae of the family Noctuidae (Hardwick 1965; Mitter *et al.* 1993). The biology and ecology of the species within this complex have recently been reviewed by Fitt (1989), Zalucki (1991) and King (1994). It is significant that within the group there exists a large number of highly destructive crop pests against which an unparalleled variety and quantity of insecticides have been used. The polyphagous nature of a number of these species, their wide geographic range and their ability to adapt to diverse cropping systems have contributed to this pest status. Moreover, the ability of certain species within the complex to develop resistance to insecticides has placed the heliothine Lepidoptera among a handful of the world's most significant crop pests.

The genus *Helicoverpa* (designated *Heliothis* for a period) includes the Old World species *Helicoverpa armigera*, generally considered to be the most important species within this group. Commonly known as the cotton bollworm, gram podborer or American bollworm, *H. armigera* occurs in Africa, Asia, southern Europe and Australia and is a major pest of cotton, maize, sorghum, pigeonpea, chickpea, soyabean, groundnut, sunflower and a range of vegetables. It, above all others in this genus, has developed resistance to virtually all of the insecticides that have been deployed against it in any quantity. *Helicoverpa punctigera* is a pest of cotton, sunflower, lucerne, soyabean, chickpea and safflower in Australia and is commonly found alongside *H. armigera*. Interestingly, until recently *H. punctigera* had not developed resistance to insecticides (Gunning & Easton 1994; Gunning *et al.* 1994), and this may have been because the pool of unsprayed insects is so vast that the treated proportion of the total population is only trivial (Forrester *et al.* 1993). Any resistance genes would be swamped by susceptible genes in the unsprayed refugia. Nevertheless, a field population of *H. punctigera* from New South Wales was recently shown to have developed resistance to a pyrethroid (Gunning *et al.* 1997). The oligophagous species *Helicoverpa assulta* feeds on tobacco and other solanaceous plants and is found in Africa, Asia, parts of Australasia and the South Pacific. It is normally considered to be a minor pest and there is no evidence of it having developed resistance anywhere within its range since it is not subject to any significant insecticide treatment (Armes *et al.* 1996). In the New World this genus is represented by the corn earworm, *Helicoverpa zea*, a key pest of maize, sorghum, cotton, tomato, sunflower and soyabean. It too has developed resistance to a number of the insecticide groups used against it (Sparks 1981; Wolfenbarger *et al.* 1981; Stadelbacher *et al.* 1990), although not all the crops it attacks are sprayed and the species remains reasonably amenable to control with insecticides.

Within the genus *Heliothis* there are two species of note. The New World representative, *Heliothis virescens*, is distributed throughout the Americas, is commonly known as the tobacco budworm and is a major pest of cotton, tobacco, tomato, sunflower and soyabean. Like *H. armigera* above, it has developed resistance to all the insecticides that have been used against it in significant quantities (Sparks 1981; Wolfenbarger *et al.* 1981; Sparks *et al.* 1993). The polyphagous species, *Heliothis peltigera*, has a broad distribution across central and southern Europe, the Canary Islands, Asia Minor and India and it is a pest of safflower, tobacco, cotton, chickpea, fodder crops, grapevines and various fruit trees. Resistance to insecticides has not been reported in this species and this is presumed to be due to lack of intense selection.

It is clear that within the Heliothinae there are two very significant pest species that have been subjected to intense selection with a range of insecticides and which have developed significant levels of resistance to insecticides: *H. armigera* and *H. virescens*. This review will therefore concentrate on resistance in these species and will attempt to compare and contrast the phenomenon, particularly with respect to the mechanisms of resistance. There are a large number of reviews that document the historical development of resistance in these species and it is not the intention of the present author to rehearse all of this literature here. The reader is directed to earlier works by Sparks (1981), Wolfenbarger *et al.* (1981) and Sparks *et al.* (1993). Most studies on resistance in heliothine insects in the past 15 years have concerned the pyrethroid insecticides and it is for this reason that this review will place special emphasis on this group, although other insecticide groups will also be considered.

An understanding of the mechanisms underlying resistance is central to an ability to continue to effectively use existing insecticide chemistry to which resistance has already developed. A knowledge of the mechanisms of resistance enables one to understand not only the cross-resistance patterns within insecticide groups but also those between them. Thus, the mechanisms of resistance determine the use of 'resistance-breaking' compounds and areas of new insecticide chemistry. Such considerations are crucial in resistance management. A detailed knowledge of resistance mechanisms could also be considered as essential in the formulation of diagnostics for use in resistance management although, as will be emphasized later, the very diversity of response to selection in these insects could make the practical use of such diagnosis especially difficult. The design of expression systems for use in insecticide discovery might also be usefully influenced by such information. This review therefore places considerable emphasis on the mechanisms of resistance and compares them in heliothine populations around the world.

2. RESISTANCE TO ORGANOPHOSPHATES

(a) Heliothis virescens

After the development of resistance to DDT and toxaphene, organophosphates (OPs), particularly methyl parathion, were introduced into the USA to control *H. virescens* and *H. zea*. Resistance developed in *H. virescens* within a few years of OP introduction (Wolfenbarger & McGarr 1970; Harris 1972), and had become widespread throughout the southern states of the USA by 1980 (Sparks *et al.* 1993). Numerous reports have detailed the progress of resistance to OPs including methyl parathion, sulprofos and profenofos both within and between seasons (e.g. Wolfenbarger 1981; Elzen *et al.* 1992; Kanga & Plapp 1992, 1995; Sparks *et al.* 1993; Graves *et al.* 1994; Kanga *et al.* 1995; Martin *et al.* 1995, 1997). Outside the USA, a low level of resistance to monocrotophos was noted in Colombia (Ernst & Dittrich 1992).

(b) Helicoverpa zea

Resistance to methyl parathion was reported in some states of the USA and Central America (Wolfenbarger *et al.* 1981), although there is little supporting information in the literature that resistance to OPs is a significant problem in the control of this species (Sparks *et al.* 1993).

(c) Heliothis armigera

H. armigera in Australia have generally been considered to be relatively susceptible to OP insecticides. Gunning & Easton (1993) found no evidence of resistance to methyl parathion and today only low levels of resistance are found to profenofos, chlorpyrifos and methyl parathion (N. W. Forrester, personal communication). In contrast, high levels of resistance to monocrotophos and low levels of resistance to chlorpyrifos and profenofos have been recorded in populations of *H. armigera* in Pakistan (Ahmad *et al.* 1995), although resistance to profenofos is continuing to rise as growers opt to use OPs rather than the pyrethroids, to which there is resistance. Low-to-moderate resistance was found to quinalphos in Indian and Pakistani populations of *H. armigera* (Armes *et al.* 1996), but there was no evidence of significant resistance to monocrotophos. Since 1980, phoxim has been the most widely used OP for the control of *H. armigera* in China. It was highly effective until 1990, when it failed to control populations in North China. Bioassays with insects collected from different geographical areas of China during 1994 and 1995 showed resistance to phoxim to be widespread (Wu *et al.* 1997). No resistance to monocrotophos was observed in 1992 and 1993 (Wu *et al.* 1995), but higher levels were recorded in 1995 (Wu *et al.* 1996). No resistance to OPs was detected in *H. armigera* in Thailand (Ahmad & McCaffery 1988).

3. MECHANISMS OF RESISTANCE TO ORGANOPHOSPHATES

(a) *Insensitive acetylcholinesterase: target-site resistance to organophosphates*

The enzyme acetylcholinesterase resides on the post-synaptic membrane of cholinergic synapses and is responsible for the breakdown of acetylcholine after stimulation of nicotinic acetylcholine receptors on the postsynaptic neuron. Both organophosphate and carbamate insecticides prevent the breakdown of acetylcholine by inhibiting the activity of this enzyme. The increased residence time of acetylcholine in the synapse causes repeated stimulation of the postsynaptic neuron and hence neuronal hyperactivity. Commonly, resistance to OPs involves the selection of mutants that possess a form of the enzyme insensitive to inhibition.

A large number of reports have shown that resistance to OPs in *H. virescens* may be due, at least in part, to a target-site resistance involving decreased sensitivity of acetylcholinesterase to inhibition (Brown & Bryson 1992; Kanga & Plapp 1995; Brown *et al.* 1996a; Harold & Ottea 1997). In resistant strains of *H. virescens*, Brown & Bryson (1992) and Gilbert *et al.* (1996) demonstrated the presence of acetylcholinesterase insensitive to inhibition by methyl paraoxon and G. Zhao *et al.* (1996) demonstrated acetylcholinesterase insensitive to paraoxon. Although this mechanism may be common in OP-resistant insects it may not be universal within field populations of *H. virescens*, as shown by Harold & Ottea (1997).

(b) *Metabolic mechanisms of resistance to organophosphates*

Metabolic resistance to organophosphate insecticides in heliothine insects has been thought to be due to elevation in the activity of number of detoxification systems. Most frequently, resistance to these insecticides has been correlated with elevated esterase activity, especially when the model substrate 1-naphthyl acetate (1-NA) is used; this result suggests a strong association between these enzymes and OP resistance. Esterase synergists such as TBPT and EPN were shown to be effective against methyl parathion resistance in *H. virescens* (Payne & Brown 1984). Importantly, in this New World species, elevated esterase activities were shown to be responsible for resistance to OPs such as methyl parathion, profenofos and azinphosmethyl and for cross-resistance between carbamate, OP and pyrethroid insecticides (Goh et al. 1995; G. Zhao et al. 1996). Higher phosphotriester hydrolase activity was reported to be involved in resistance to methyl parathion in *H. virescens* from North Carolina (Konno et al. 1989). In a recent study, high frequencies of profenofos resistance were recorded in larvae of all of a number of field strains of *H. virescens* collected from Louisiana in 1995 and were strongly correlated with esterase activity (Harold & Ottea 1997).

Glutathione S-transferases have also been frequently associated with resistance to OPs and thought to be responsible for metabolism of these compounds (Whitten & Bull 1978). Resistance to profenofos was shown to be only moderately correlated with glutathione S-transferase activity towards 1-chloro-2,4-dinitrobenzene (CDNB) and had no correlation with glutathione S-transferase activity to 1,2-dichloro-4-nitrobenzene (DCNB) (Ibrahim & Ottea 1995; Harold & Ottea 1997). This correlation of profenofos resistance with activity of GST towards CDNB but not DCNB suggests that these GST enzymes have different identities and therefore likely contributions to profenofos resistance. No differences in GST activity were observed by Konno et al. (1989).

Metabolism of OP insecticides by P450 monooxygenases was reported in a number of early studies (Whitten & Bull 1974; Reed 1974; Brown 1981; Bull 1981). Martin et al. (1995) identified low to moderate levels of resistance to profenofos and sulprofos in Louisiana, Mississippi and Texas and later showed that profenofos was synergized by PBO in populations of *H. virescens* from Texas, Mississippi and Oklahoma (Martin et al. 1997). Other research suggested that monooxygenases might not be involved in the elimination of OPs (Gould & Hodgson 1980; Payne & Brown 1984; Konno et al. 1989). Most recently, Harold & Ottea (1997) found no correlation between profenofos resistance and P450 monooxygenase activity towards the model substrate *p*-nitroanisole.

4. RESISTANCE TO CARBAMATES

(a) Heliothis virescens

Resistance to the oxime carbamates thiodicarb and methomyl has been recorded a number of times in populations of *H. virescens* from Louisiana, Mississippi, Texas and Arkansas (Sparks 1981; Elzen et al. 1992; Martin et al. 1992, 1995; Sparks et al. 1993; Kanga & Plapp 1995), and also in populations from Mexico (Roush & Wolfenbarger 1985).

(b) Helicoverpa zea

There are no reports of significant resistance to carbamates in *H. zea*.

(c) Helicoverpa armigera

Thiodicarb and methomyl are the carbamates most widely used against *H. armigera* in Australia. Methomyl resistance was noted in 1986 but the insect remained susceptible to thiodicarb for a number of years more (Gunning et al. 1992). Resistance to thiodicarb was detected in New South Wales in 1993 and this gave cross-resistance to methomyl (Gunning et al. 1996b). Since then resistance to thiodicarb has increased and moderate resistance to carbamates is now common (N. W. Forrester, personal communication). In China, significant resistance to methomyl was recorded in strains of *H. armigera* from Shandong province (Wu et al. 1995, 1996). Low-level resistance to thiodicarb was seen in *H. armigera* from Pakistan (Ahmad et al. 1995). Substantial resistance to methomyl was recorded in populations from cotton-growing areas of Andhra Pradesh, India (Armes et al. 1992, 1996), with lower levels being more typical of other locations including Nepal, Gujarat and Maharashtra.

5. MECHANISMS OF RESISTANCE TO CARBAMATES

(a) *Insensitive acetylcholinesterase: target-site resistance to carbamates*

Target-site resistance to carbamates is similar to that found with organophosphates (see above). Acetylcholinesterase insensitive to inhibition by propoxur and methomyl was observed in a selected strain of *H. virescens* (Brown & Bryson 1992). More recently, insensitive acetylcholinesterase was shown to be a major mechanism of resistance to methomyl and carbaryl in strains of *H. virescens* and to thiodicarb in a thiodicarb- and pyrethroid-resistant strain (G. Zhao et al. 1996). In Australia the recently developed resistance to thiodicarb in *H. armigera* has been shown to be due to a form of acetylcholinesterase that is insensitive to both thiodicarb and methomyl.

(b) *Metabolic resistance to carbamates*

Both enhanced esterase and enhanced monooxygenase activity have been found to be significant mechanisms of resistance to carbamates. In one recent study, substantially increased esterase activity was observed and thought to be responsible for resistance in a thiodicarb-resistant (and pyrethroid-resistant) strain of *H. virescens* (Goh et al. 1995). Rose et al. (1995), using a similar strain, found high levels of P450 monooxygenase activity as well as increased esterase activity. The involvement of P450 monooxygenases was considered likely by G. Zhao et al. (1996), who showed significant synergism of thiodicarb with PBO. They also inferred the involvement of enhanced esterase activity in this resistance. Very recently, PBO was shown to synergize the action of methomyl and thiodicarb in a number of field strains although it antagonized the action of thiodicarb in some strains (Martin et al. 1997).

6. RESISTANCE TO CYCLODIENES

(a) Heliothis virescens

Resistance to endosulfan has been demonstrated in strains of *H. virescens* from Louisiana, Mississippi, Texas and Arkansas (Elzen *et al.* 1992; Kanga *et al.* 1995; Martin *et al.* 1995).

(b) Helicoverpa zea

Increased tolerance to endosulfan was found in field populations of *H. zea* from Texas in 1994 (Kanga *et al.* 1996). There appear to be no other records of significant resistance to endosulfan in this species.

(c) Helicoverpa armigera

Resistance to endosulfan in *H. armigera* has recorded in Australia since the early 1970s and a number of reports have demonstrated the substantial and continuing nature of this problem (Kay 1977; Forrester *et al.* 1993; Gunning & Easton 1994). Current levels of resistance to endosulfan in Australia are moderate. Relatively low levels of resistance were characteristic of *H. armigera* in various regions of India from 1988 to 1992 (McCaffery *et al.* 1989; Armes *et al.* 1992). Rather higher levels of resistance to this compound were found in later years by Armes *et al.* (1996), who suggested that incipient resistance to endosulfan was present in this species in India, Nepal and Pakistan. Low resistance to endosulfan characterized populations of *H. armigera* from Pakistan between 1991 and 1993, but thereafter resistance rose to peak frequencies in 1995, falling back somewhat in later years (Ahmad *et al.* 1995, 1998). Populations collected from Indonesia in 1987 and 1988 were reported to be resistant to pyrethoids (McCaffery *et al.* 1991a).

7. MECHANISMS OF RESISTANCE TO CYCLODIENES

(a) Altered GABA receptor: target-site resistance to cyclodienes

The GABA-gated chloride-ion channel receptor complex is generally considered to be the target for cyclodiene insecticides such as endosulfan. These compounds act as GABA antagonists and hence, because they suppress the inhibitory transmitter action of GABA, their action results in increased postsynaptic neuronal activity. Although no direct evidence has been obtained with heliothines, target-site insensitivity to cyclodiene action has been inferred in adult *H. virescens* on the basis of highly correlated toxicities of dieldrin and endosulfan (Kanga & Plapp 1995).

8. RESISTANCE TO PYRETHROID INSECTICIDES

(a) Introduction

The pyrethroid insecticides were introduced to replace the resistance-prone and environmentally unsuitable organochlorines (OCs), cyclodienes and organophosphates (OPs) (Morton & Collins 1989). They clearly had a number of distinct advantages over insecticides used previously. They possessed an inherently high activity and could be applied at extremely low doses for the control of a huge range of public health and agricultural pests. Their high activity meant that effective foliar profiles were maintained for considerable periods. They were safe to mammals, had low environmental impact and were immobile in the soil (Elliott 1989). The pyrethroids were especially useful in cotton, where their contact activity and good efficacy enabled the grower to

regain control of pest species that had become resistant to previously used insecticides. The global demise of the effectiveness of pyrethroids has provoked a huge research effort directed at understanding the nature of this resistance and hence alternative control strategies.

(b) Resistance to pyrethroid insecticides around the world

(i) Heliothis virescens *in the USA*

Following the development of resistance to DDT, methyl parathion and a growing number of other OPs (Sparks *et al.* 1993), the pyrethroid insecticides were introduced into the USA and became available for use on cotton in 1978, quickly becoming the insecticides of choice. A small number of studies had inferred a degree of cross-resistance to pyrethroids in methyl parathion-resistant strains of the tobacco budworm, although analysis of these data revealed no significant trends. Nevertheless, susceptibility to pyrethroids was correlated with that to methyl parathion (Sparks *et al.* 1993), and suggested that differences in susceptibility were already present in populations of the tobacco budworm in cotton.

Numerous studies have documented resistance to pyrethroids in *H. virescens* in the USA and the reader is directed to the comprehensive review by Sparks *et al.* (1993) for more details. Although significant changes in susceptibility had been noted in the Imperial Valley of California in the early 1980s (Twine & Reynolds 1980; Martinez-Carrillo & Reynolds 1983), these were not considered to have led to any field failure. The first reports of significant resistance appeared in 1985 in west Texas (Plapp & Campanhola 1986) and these were quickly followed by a range of similar findings throughout the cotton-belt states of the southern USA, including Alabama (Mullins *et al.* 1991), Arkansas (Plapp *et al.* 1987, 1990), Louisiana (Leonard *et al.* 1988; Plapp *et al.* 1990; Elzen *et al.* 1992), Mississippi (Luttrell *et al.* 1987; Plapp *et al.* 1990; Elzen *et al.* 1992; Ernst & Dittrich 1992), Oklahoma (Plapp *et al.* 1990) and Texas (Plapp *et al.* 1987, 1990). In many cases this resulted in considerable cross-resistance between pyrethroids and this was thought to imply the presence of a target-site mechanism of resistance (Martin *et al.* 1992; Graves *et al.* 1993; Sparks *et al.* 1993). Because a complete loss of pyrethroids was feared, resistance monitoring programmes were instituted (Plapp *et al.* 1987), and management plans organized in Texas and the mid-south in an effort to provide pyrethroid-free windows during the cotton-growing season (Sparks *et al.* 1993). Interestingly, the continued used of pyrethroids in the USA has led to what appears to be a shift in the mechanisms of resistance to pyrethroids, as detailed below.

(ii) Heliothis virescens *in Mexico*

H. virescens is a common pest of cotton in Mexico and pyrethroids have been extensively used for its control since the early 1980s. Monitoring for resistance to pyrethroids has been conducted in agricultural regions of northwestern Mexico since 1984, when resistance was first noted. High levels of resistance were recorded in 1987 from populations from the Yaqui and Mexicali valleys and in the 1988 season from the Costa de Hermosillo and Region de Caborca (Martinez-Carrillo 1991, 1995). These high levels of resistance prompted the introduction, in 1989, of a strategy to reduce pyrethroid selection pressure in the Yaqui Valley. As a result, pyrethroid resistance decreased in this area in 1988 and 1989 and has remained stable since 1990 (Martinez-Carrillo 1995). In contrast, levels of resistance in the northeast of the country are high and would be expected to cause control problems.

(iii) Heliothis virescens *in Colombia*

Pyrethroids became available for use in cotton in the late 1970s and early 1980s and were very extensively used, to the exclusion of other products. Very substantial resistance to cypermethrin in the tobacco budworm was noted from 1985 and has been documented by Ernst & Dittrich (1992) and confirmed by McCaffery (1994).

(iv) Helicoverpa zea *in the Americas*

Very extensive resistance to DDT was a feature of early control of *H. zea* in the USA (see, for example, Graves *et al.* 1963; Wolfenbarger *et al.* 1981; Sparks *et al.* 1993). The first substantial report of resistance to pyrethroids in *H. zea* was that of Stadelbacher *et al.* (1990). Following this, a number of other authors noted a loss of susceptibility to pyrethroids in this species (Graves *et al.* 1993; Abd-Elghafar *et al.* 1993; Kanga *et al.* 1996; Bagwell *et al.* 1997). Despite this loss of susceptibility, pyrethroid insecticides presently remain effective for the control of *H. zea* in US cotton, even at low field application rates. In one of the few studies on this species conducted outside the USA, strains of *H. zea* from the Tiquisate area of Guatemala and the Leon area of Nicaragua were found to be very substantially resistant to cypermethrin (Ernst & Dittrich 1992).

(v) Helicoverpa armigera *in Australia*

Before the introduction of pyrethroids in 1977 in Australia, *H. armigera* had developed severe resistance to DDT in the Ord River Valley (Wilson 1974), New South Wales (Goodyer *et al.* 1975; Goodyer & Greenup 1980) and Queensland (Kay 1977). Resistance to endosulfan (Kay 1977; Kay *et al.* 1983; Gunning & Easton 1994), OPs (Goodyer & Greenup 1980; Kay *et al.* 1983) and carbamates (Gunning *et al.* 1992) was also known to be present. Resistance to pyrethroids first appeared in 1983 (Gunning *et al.* 1984), and immediately a resistance management strategy was implemented, which restricted the use of pyrethroids to a 42-day window during January–February (from 1990 they were restricted to a 35-day window) (Forrester 1990; Forrester *et al.* 1993). Endosulfan use was also limited. An effective weekly monitoring scheme based on survival of fourth-instar larvae of *H. armigera* after treatment with a diagnostic dose of fenvalerate was initiated and much data accumulated on the effects of selection and survival of resistant individuals. Later monitoring also determined the likely presence of a metabolic resistance based on enhanced monooxygenase activity by treating larvae with both fenvalerate and the metabolic inhibitor piperonyl butoxide (PBO) (see below). Based on these results, PBO could be added to the last of the three (maximum) sprays in the pyrethroid window. This strategy undoubtedly held pyrethroid resistance in check for a number of years although there appeared to be a steady rise in the proportion of the population that was resistant to pyrethroids (Forrester *et*

al. 1993). *H. armigera* in unsprayed refugia readily became contaminated with resistant individuals (Gunning & Easton 1989; Forrester *et al.* 1993), and similar levels of resistance were found in other crops, such as maize (Glenn *et al.* 1994). This gradual loss of pyrethroid efficacy together with the development of an immunodiagnostic to distinguish the eggs of *H. armigera* from those of *H. punctigera*, the use of *Bacillus thuringiensis* (*Bt*) and other insecticides and the advent of *Bt*-transgenic cotton led to a complete reorganization of the strategy and a relaxation on the use of the now less useful pyrethroids. The situation is continuing to deteriorate, with resistance to pyrethroids increasing steadily (N. W. Forrester, personal communication).

(vi) Helicoverpa armigera *in New Zealand*

A programme to monitor resistance to fenvalerate in *H. armigera* was initiated in 1991 in tomato, maize and lucerne crops in New Zealand. A significant trend of declining mortality from 1992 to 1994 was seen and this suggests an increase in the frequency of resistance to the pyrethroids (Cameron *et al.* 1995; Suckling 1996). Management strategies have been devised to counter this problem (Suckling 1996).

(vii) Helicoverpa armigera *in Thailand*

Wangboonkong (1981) first reported inadequate control of *H. armigera* in Thailand soon after the introduction of pyrethroids, but it was not known whether resistance was the cause. Significant resistance to pyrethroids was found in populations of *H. armigera* from the Tak Fa area of Nakonsawan in Thailand in 1985 (Ahmad & McCaffery 1988). These insects were also resistant to DDT and carbaryl. Pyrethroid resistance was again noted in Thai populations of the insect by Ernst & Dittrich (1992).

(viii) Helicoverpa armigera *in Indonesia*

After the introduction of pyrethroids in the 1980s, resistance to was found in populations of *H. armigera* collected from the cotton-growing areas of South Sulawesi, Indonesia, in 1987 and early 1988 (McCaffery *et al.* 1991a). These populations were also resistant to endosulfan and DDT.

(ix) Helicoverpa armigera *in China*

Almost all groups of conventional insecticides have been used to control *H. armigera* in China. DDT resistance was first detected in *H. armigera* in Henan province (Anon 1974), and subsequently in Jiangsu and Hebie provinces (Zhu *et al.* 1982), together with resistance to carbaryl. Pyrethroids such as fenvalerate and deltamethrin have been widely used since 1983 with others such as cyhalothrin, cypermethrin, esfenvalerate, fenpropathrin and cyfluthrin being used from the mid- to late-1980s. There were no substantial changes in susceptibility until around 1989, but in the following years resistance to pyrethroids was widely detected in a number of areas including Jiangsu, Henan and Shandong provinces (Tan *et al.* 1987; Shen *et al.* 1991, 1992, 1993; Wu *et al.* 1996, 1997b). The development of this resistance led to calls for a resistance management strategy to restrict pyrethroid use, to promote greater emphasis on the use of alternations with other insecticides and to promote the use of biological control (Shen *et al.* 1992). Although levels of resistance to pyrethroids are still high, recent lower populations have alleviated the problem to some degree (Y. Wu, personal communication).

(x) Helicoverpa armigera *in Central Asia*

High levels of resistance to pyrethroids (as well as to OCs and OPs) have been found in *H. armigera* from Tajikstan and Azerbaijan (Sukhoruchenko 1996). In a similar study, resistance to pyrethroids was found to be present in populations of *H. armigera* from Russia (Leonova & Slynko 1996).

(xi) Helicoverpa armigera *in India*

Pyrethroid insecticides were first used in India in 1980 for the control of a number of pests, including *H. armigera*. In 1987 resistance to pyrethroids was first noted in India in Andhra Pradesh (Dhingra *et al.* 1988; McCaffery *et al.* 1988, 1989; Phokela *et al.* 1989) in populations that were also resistant to DDT and slightly resistant to endosulfan (McCaffery *et al.* 1989). Numerous other studies confirmed the high incidence of pyrethroid resistance, especially in the cotton- and pulse-growing regions of central and southern India, and also confirmed its gradual spread to other regions of the country (see, for example, Phokela *et al.* 1990; Mehrohtra & Phokela 1992; Armes *et al.* 1992, 1996; Sekhar *et al.* 1996; Jadhav & Armes 1996). Pyrethroid resistance has recently been found in the Punjab close to populations over the border in Pakistan, leading Armes *et al.* (1996) to the conclusion that pyrethroid resistance is ubiquitous in *H. armigera* in the ·Indian subcontinent. Resistance to pyrethroids is frequently accompanied by resistance to endosulfan, to OPs such as quinalphos and monocrotophos, and to the oxime carbamate methomyl (Armes *et al.* 1992, 1996).

(xii) Helicoverpa armigera *in Pakistan*

As a result of pyrethroid use since the early 1980s, moderate to high levels of resistance to pyrethroids were found in populations of *H. armigera* collected from various regions of Pakistan from 1991 onwards (Ahmad *et al.* 1995). These insects were also resistant to the OP monocrotophos, showed moderate resistance to endosulfan and had low-level resistance to the OPs chlorpyrifos and profenofos and the carbamate thiodicarb. Interestingly, in a subsequent study these authors showed variations in resistance to pyrethroids depending on their structure. Although resistance varied from location to location, the general trend was for moderate to high resistance to chemicals like cypermethrin, a low-to-moderate resistance to compounds like deltamethrin and comparatively low resistance to others like lambda-cyhalothrin (Ahmad *et al.* 1997). With the loss of efficacy of the pyrethroids farmers have begun to use other non-pyrethroid compounds, with the result that levels of pyrethroid resistance were lower in 1997 than in previous years (Ahmad 1998).

(xiii) Helicoverpa armigera *in Africa*

In the Ivory Coast pyrethroids have been applied for 15 years to control *H. armigera* and other bollworms. These pyrethroids were always mixed or rotated with organophosphate insecticides in an effort to prevent or delay resistance in bollworms (Alaux *et al.* 1997). Ernst & Dittrich (1992), in a comparative survey of resistance in heliothines around the world, could find no evidence for resistance to pyrethroids in the Ivory Coast. Vassal *et al.* (1997) confirmed that before 1992 there was no change in resistance to pyrethroids but in subsequent years susceptibility decreased and by 1995 and 1996 significant resistance was recorded. This is the first documented evidence for resistance to pyrethroids in bollworms in West Africa. No resistance to pyrethroids was found in populations of *H. armigera* from Chad, although some changes in tolerance were believed to be occurring (Martin & Renou 1995).

(xiv) Helicoverpa armigera *in Turkey*

Resistance to synthetic pyrethroids was found in populations of *H. armigera* in 1984, after their initial use around 1980 (Anon 1986). Similar findings were reported by Ernst & Dittrich (1992).

(xv) Helicoverpa armigera *in Israel*

Since 1987 a strictly observed insecticide resistance management strategy has been in place in cotton fields in Israel. This is designed to maintain susceptibility to a range of insecticides, including pyrethroids, in *H. armigera* and other cotton pests. Monitoring studies show that, despite slight fluctuations during the season, susceptibility to cypermethrin did not alter during the period 1987–1991 (Horowitz *et al.* 1993); control continued to be achieved despite a very marked decline in the number of sprays applied (Horowitz *et al.* 1995).

(xvi) Helicoverpa punctigera *in Australia*

A population of *H. punctigera* collected from New South Wales in 1994 was shown to be resistant to fenvalerate (Gunning *et al.* 1997). This is the first report of significant resistance in this species.

9. MECHANISMS OF RESISTANCE TO PYRETHROIDS

(a) *Nerve insensitivity: target-site resistance to pyrethroids*

The principal site of action of DDT and pyrethroids is the voltage-gated sodium channel of nerve cells (Soderlund & Bloomquist 1989; Narahashi 1992; Bloomquist 1996). These insecticides alter the gating kinetics of the sodium channel so that the open time of the channels is increased after the passage of the depolarizing pulse of an action potential. This inhibition of sodium-channel inactivation leads to the development of prolonged sodium currents and accounts for the prolonged depolarizing after-potential. This action causes the repetitive firing of neurons that is typically found in pyrethroid-poisoned insects. Pyrethroids also cause membrane depolarization due to the prolonged opening of sodium channels. Type II pyrethroids, which contain a cyano group at the α position, are generally more potent in this respect than type I pyrethroids, which lack this α-cyano group. Thus sensory neurons are stimulated as a result of membrane depolarization. Membrane depolarization at nerve terminals causes massive release of neurotransmitter, resulting in severe disruption of synaptic transmission.

(i) *Indirect evidence for nerve insensitivity*

Evidence for the involvement of a *kdr*-like mechanism of resistance in heliothine insects has not been easy to obtain. There is much indirect evidence that implies the involvement of a target-site mechanism of resistance to DDT and pyrethroids, although such evidence is never wholly reliable. The most frequently used criterion has been the lack of synergizable resistance. Weekly estimates of the survival of third-instar *H. armigera* treated with a discriminating dose of fenvalerate both with and without PBO have been used in Australia to provide an estimate of the percentage of non-synergizable resistance (Forrester *et al.* 1993). It is inferred that this residual resistance is due to other mechanisms, including a target-site resistance. Similar findings were presented by Armes *et al.* (1996) working with Indian populations of *H. armigera*. In a study in Maharashtra State in India, it was suggested that the residual non-metabolic resistance remaining after synergism with both the monooxygenase synergist piperonyl butoxide (PBO) and the esterase synergist *S,S,S,*-tributyl phosphorotrithioate (DEF) was likely to be due to target-site insensitivity (Kranthi *et al.* 1997), and this is being verified now. Given that the principal mechanisms of resistance in these insects are considered to be enhanced monooxygenase activity and/or esterase activity and a target-site resistance of the *kdr* type, this approach might appear reasonable. Nevertheless, as indicated below, PBO may not be a reliable synergist for monooxygenases and it may indeed synergize other forms of metabolic resistance as suggested by Gunning *et al.* (1996a). Moreover, non-synergizable penetration resistance also contributes to this residual resistance. Given such considerations the premise that non-synergizable resistance represents target-site resistance is at best equivocal.

The presence of cross-resistance between DDT and pyrethroids is frequently used as evidence for the involvement of resistance at the target site. *H. armigera* from Thailand were shown to possess high levels of resistance to DDT and cypermethrin; this observation implies a common mechanism. Lack of synergism by the DDT-dehydrochlorinase synergist FDMC reinforced this view (Ahmad & McCaffery 1991); as discussed below, the insects were indeed shown to possess a nerve insensitivity (Ahmad *et al.* 1989). Recent studies with strains of *H. armigera* selected from field collections from Jiangsu province in China have shown highly significant, non-synergizable cross-resistance between DDT and fenvalerate (J. Tan & A. R. McCaffery, unpublished results). This resistance has subsequently been shown to be due to nerve insensitivity and its molecular basis is being studied presently (see below). The inability to identify metabolites of pyrethroids in biochemical studies of pyrethroid metabolism has also been used to imply that resistance may be due to target-site resistance, although it is clear that the common involvement of both metabolic and non-metabolic mechanisms in the same individuals in resistant strains makes such an approach difficult.

Finally, because individuals with target-site resistance might theoretically be able to withstand higher internal concentrations of insecticide than their susceptible counterparts, it has been thought that the presence of high titres in insects that survive such treatments indicates the presence of this mechanism. The nervous system of resistant third-instar larvae of the PEG87 strain of *H. virescens* was shown to contain up to tenfold greater concentrations of *cis*-cypermethrin than those of susceptible larvae of the BRC strain (Wilkinson &

McCaffery 1991). In addition, the behavioural responses of these intoxicated insects suggested that comparable symptoms of intoxication occurred at higher concentrations in larvae of the resistant strain than in larvae of the susceptible strain. This was again taken to imply a decreased interaction of the pyrethroid with its target site. At best, such evidence is tenuous. Such considerations form the basis of behavioural assays for nerve insensitivity, typified by the hot-needle assay developed by Bloomquist & Miller (1985) and a locomotory assay developed by Gunning (1996).

(ii) *Direct evidence for nerve insensitivity*

There now exists a large body of direct evidence that a form of nerve insensitivity contributes substantially to many cases of resistance to DDT and pyrethroids in heliothine insects. Nicholson & Miller (1985) first demonstrated this neurophysiologically in a resistant strain of *H. virescens* collected from cotton-growing areas of southern California. A similar technique was used to demonstrate nerve insensitivity in a pyrethroid- and DDT-resistant Thai strain of *H. armigera* (Ahmad *et al.* 1989). At the onset of pyrethroid resistance in Australia in 1983 a strong *super-kdr*-like mechanism was demonstrated by using a simple single-dose neurophysiological technique (Gunning *et al.* 1991), but in a survey during the period from 1997 to 1990 no evidence was found for the presence of this *super-kdr*-like mechanism. Instead, another distinct *kdr*-type mechanism with little or no toxicological significance was found. By means of a cumulative dose–response neurophysiological assay for spontaneous neuronal activity, nerve insensitivity to cypermethin was demonstrated in resistant laboratory strains of *H. virescens* (Gladwell *et al.* 1990) and in field strains collected from various parts of the US cotton belt (McCaffery *et al.* 1995; Ottea *et al.* 1995). Since monitoring of pyrethroid resistance in the USA has been based upon the adult vial test (Plapp *et al.* 1987), it is significant that nerve insensitivity in adult stages of resistant strains of *H. virescens* was correlated with that in larval stages (Holloway & McCaffery 1996). Modifications of this technique have also been used to demonstrate high levels of nerve insensitivity to pyrethroids and DDT in *H. armigera* from Andhra Pradesh state in India (West & McCaffery 1992) and from various parts of China (Y. Zhao *et al.* 1996; McCaffery *et al.* 1997; Ru *et al.* 1997; Zhang *et al.* 1997), and *H. zea* from the USA (Holloway *et al.* 1997).

(iii) *Molecular basis of nerve insensitivity resistance to pyrethroids*

Although pyrethroids may interact with a number of sites within the nervous system and although a range of effects may be produced by these interactions, the principal site of action is considered to be the voltage-gated sodium channel. For this reason efforts to determine the molecular basis of resistance to pyrethroids in heliothine insects have centred on changes in sodium channels and have followed similar pioneering studies on house flies and cockroaches. Experiments conducted by Church & Knowles (1992) on binding to neural membranes of saxitoxin, a high-affinity neurotoxin binding to site 1 on the sodium channel, suggest that there is no difference in the number of sodium channels between pyrethroid-resistant and -susceptible strains of *H. virescens*. Further work has shown that binding of batrachotoxin, a sodium-channel neurotoxin, is enhanced by pyrethroid binding; by means of this assay these authors have provided evidence that the affinity for pyrethroids on the sodium channels is considerably reduced in resistant *H. virescens* compared with susceptible counterparts (Church & Knowles 1993). Taken together these studies imply that reduced affinity of binding is responsible for resistance to pyrethroids at the sodium channel in this species.

The evidence reviewed above suggested that resistance to pyrethroids and DDT might be expected to result from the selection of genetic mutants with altered sodium channels. Molecular genetic studies on sodium channels would clearly be essential to understand the basis of this resistance. By using degenerate oligonucleotide primers based on conserved amino-acid sequences in sodium channels of *Drosophila melanogaster*, *para*-homologous sodium-channel genes were isolated from a range of insects including *H. virescens* (Doyle & Knipple 1991). The polymerase chain reaction was used to amplify sequences from genomic DNA from the PEG87 strain of *H. virescens* by using degenerate primers homologous to the fourth transmembrane domain of the α-subunit locus *para* of *D. melanogaster* (Taylor *et al.* 1993). One genomic clone encoding a putative sodium channel in *H. virescens* was obtained and designated *hscp* (*Heliothis sodium channel para* homologue). In a subsequent experimental analysis, markers for *hscp* were found to be linked to resistance phenotypes and this provided the first molecular genetic evidence for such a link in any heliothine.

Sequence comparisons between resistant and susceptible genotypes of house fly have revealed the presence of a single leucine-to-phenylalanine mutation (L1014F) in transmembrane segment 6 of domain II associated with *kdr* resistance, and an additional methionine-to-threonine mutation (M918T) associated with *super-kdr* resistance (Williamson *et al.* 1996). Park & Taylor (1997) examined *H. virescens* in a similar manner and revealed the existence of a leucine-to-histidine change (L1029H) associated with resistance to pyrethroids and located at a position homologous to that in *kdr* strains of the house fly. No mutation homologous to that found in *super-kdr* flies was found in *H. virescens* (Park & Taylor 1997). Interestingly, the resistant PEG87 strain of this insect was not found to carry this mutation; this observation leads to the suggestion that more than one sodium-channel mutation may be contributing to pyrethroid resistance in field populations of *H. virescens*. This contrasts with the situation in *Musca domestica*, *Blattella germanica*, the diamondback moth, *Plutella xylostella*, and the peach–potato aphid, *Myzus persicae*, in which the leucine-to-phenylalanine substitution is always consistently present in resistant genotypes (Martinez-Torres *et al.* 1997). More recently, Park *et al.* (1997) reported a valine-to-methionine (V421M) substitution in transmembrane segment 6 of domain I (IS6) of the *hscp* locus of individuals of the homozygous resistant strain used by Taylor *et al.* (1993) for linkage analysis. More recently still, Head *et al.* (1998) made sequence comparisons between resistant and susceptible strains of both *H. virescens* and *H. armigera* and showed consistent aspartic acid-to-valine (D1561V) and glutamic acid-to-glycine (E1565G) substitutions in the cytoplasmic linker region between domains III and IV (III–IV) of the

para-homologous sodium-channel sequence of neurophysiologically resistant insects of both species; this region is involved in channel inactivation. A further mutation in the IIS5–IIS6 linker region was again consistently found in both resistant *H. armigera* and resistant *H. virescens* (Head 1998). All these findings emphasize the likelihood that a number of possible mutations can confer resistance at the sodium channel, although the function of these mutations clearly remains to be ascertained. The development of diagnostic technologies based on mutations that unequivocally indicate the *kdr*-like nerve insensitivity resistance to pyrethroids and DDT is a clear aim of such studies. The successful deployment of technology of this type would provide a degree of precision and refinement that has so far been lacking in the monitoring of resistance-gene frequency in heliothine pests.

(b) *Metabolic mechanisms of resistance to pyrethroids*

Studies on the metabolism of pyrethroids in heliothine insects have been characterized by a degree of contradiction, which has centred on the relative roles of the principal systems of enzymic detoxication: oxidation by the microsomal P450-dependent monooxygenases (or mixed-function oxidases) and hydrolysis by esterases. Glutathione S-transferases do not appear to be involved in resistance to pyrethroids. The traditional use of synergists to give preliminary indications of the type of metabolism involved in resistance has been critically questioned in relation to resistance to pyrethroids in view of findings, discussed below, which suggest that specific synergists are no longer (or possibly never were) effective at suppressing the enzyme systems with which they have traditionally been associated. Other studies suggest that some synergists inhibit enzyme systems alternative to those with which they are normally associated. The use of model substrates is also an area of some uncertainty: the isozymes responsible for detoxication of specific insecticides may not necessarily be those involved in model substrate metabolism. The latest and most comprehensive view of this field would suggest that both oxidative and hydrolytic activity is involved in resistance to pyrethroids in heliothines and that, indeed, these species seem likely to be able to use both types of metabolism in response to appropriate selection. Such a view has considerable implications for resistance management.

(i) *Metabolic resistance in* Heliothis virescens

Initial studies on metabolism of pyrethroids in *H. virescens* suggested that monooxygenases were involved in tolerance to *trans*-permethrin (Bigley & Plapp 1978). In field strains of this insect collected from the Imperial Valley in California, enhanced metabolism of *trans*-permethrin was a shown to be a mechanism of resistance (Nicholson & Miller 1985) and it was thought that this was likely to be due to oxidative hydroxylation. Dowd *et al.* (1987) brought insects from this location into the laboratory and selected them with flucythrinate. In contrast to the findings above, they demonstrated both a qualitative and a quantitative enhancement in the ability of larvae to hydrolyse pyrethroids compared with a susceptible strain.

The PEG87 strain of *H. virescens* has been used extensively in research on resistance to pyrethroids in this species and was derived from the US83 strain, itself assembled from a series of 19 collections across the US cotton belt where control with pyrethroids had become increasingly difficult. Despite effectively being a laboratory strain, it was considered to possess the mechanisms of resistance most likely to be representative of those in the field. By using this strain it was shown that resistance to *trans*-cypermethrin and *cis*-cypermethrin was largely due to a PBO-synergizable monooxygenase, which resulted in hydroxylation of the pyrethroid in the 4'; and 2'; positions on the phenoxybenzyl moiety (Lee *et al.* 1989: Little *et al.* 1989; Clarke *et al.* 1990; McCaffery *et al.* 1991c) and later elimination of conjugated metabolites. Further studies on this mechanism showed that the resistant strain possessed a sixfold greater quantity of total cytochrome P450 and a fourfold greater quantity of cytochrome P450 reductase than did the comparable susceptible strain (Clarke *et al.* 1990). Activity was shown to be NADPH-dependent and PBO-suppressible. Significantly, it was shown in these studies that the major hydroxy-metabolites were likely to be better substrates for hydrolysis than the parent compound. This finding was considered to explain the PBO-suppressible, NADPH-dependent appearance of acid metabolites, although carboxylesterase action was considered to play a minor role in the direct hydroxylation of the pyrethroid (Clarke *et al.* 1990). Using the PEG87 strain of *H. virescens* Abd-Elghafar *et al.* (1994) presented similar evidence for oxidative metabolism of fenvalerate.

Despite the early demonstration of enhanced metabolism in the field strain from California noted above (Nicholson & Miller 1985; Dowd *et al.* 1987), metabolic resistance was considered to be rare or absent in field populations of *H. virescens* for many years. Accordingly, a number of studies showed that pyrethroid resistance was not synergized by PBO or DEF (McCaffery *et al.* 1991b; Clower *et al.* 1992), and it was considered that the majority of the resistance was likely to be due to target-site resistance of the *kdr* type. With continued use of pyrethroids, evidence for PBO-synergizable resistance began to appear in the early 1990s in various US cotton-belt states including Louisiana, Mississippi, Texas and Oklahoma (Graves *et al.* 1991; McCaffery & Holloway 1992; Elzen *et al.* 1993; Kirby *et al.* 1994; Martin *et al.* 1994, 1997; G. Zhao *et al.* 1996), suggesting a widespread and growing resistance problem based on enhanced monooxygenase activity as had been found with *H. armigera*. The existence of enhanced resistance to cypermethrin through selection with the oxime carbamate thiodicarb, and the existence of PBO synergism of these insecticides, was strongly indicative of the involvement of oxidative metabolism (G. Zhao *et al.* 1996).

A number of other findings suggest that this might not be wholly representative of the status of this mechanism. Martin *et al.* (1997) showed that application of PBO delayed penetration of pyrethroids and suggested that PBO could influence toxicity in other ways. Some strains of *H. virescens* believed to possess enhanced monooxygenase activity were shown to be entirely unresponsive to the action of PBO and instead were synergized by propynyl ethers such as TCPB (Brown *et al.* 1996b) (see below). Nevertheless, many later biochemical studies demonstrated the importance of oxidative attack in resistance to pyrethroids in *H. virescens* (Ottea *et al.* 1995;

Ibrahim & Ottea 1995; G. Zhao *et al.* 1996). In a recent study with the metabolically blocked pyrethroid fenfluthrin, a number of other structurally modified pyrethroids and several synergists Shan *et al.* (1997) confirmed that P450 monooxygenases were associated with pyrethroid resistance in this species. Using a pyrethroid- (and thiodicarb-) resistant strain of *H. virescens* strain originally collected from fields in Louisiana where field failures with cypermethrin and thiodicarb had been recorded, Rose *et al.* (1995) examined monooxygenase, esterase and glutathione S-transferase activity. Up to 4.4-fold higher quantities of cytochrome P450 were found in the gut, fat body and carcass of the resistant strain than in those of the susceptible strain and it was thought likely that these increased P450 levels represented the sum of several P450 isozymes, each of which may possess specific yet overlapping substrate specificities. Esterases and transferases were thought to be less important in conferring resistance in this strain, although transferases may be important in the production of conjugates, which form the bulk of excreted metabolites in monooxygenase-resistant *H. virescens* (Little *et al.* 1989). Interestingly, Rose *et al.* (1995) obtained incomplete synergism with PBO in this strain and offered the suggestion that isozymes involved in pyrethroid resistance might be made unresponsive to synergists by selection pressure with PBO or other insecticides, as appears to be the case in other insects. Incomplete synergism with PBO was also obtained in another *H. virescens* strain derived from field collections in Louisiana (Shan *et al.* 1997) although it could be completely synergized by the propynyl ether TCPB. The effectiveness of TCPB as a monooxygenase synergist for pyrethroid resistance in *H. virescens* was first shown by Brown *et al.* (1996b), who concluded that different classes of P450 monooxygenases were involved in resistance-associated metabolism of pyrethroids. Such a finding casts considerable doubt on the validity of previous synergism studies using PBO; the absence of PBO synergism should perhaps not be taken as an indication of the absence of enhanced oxidative metabolism.

The doubts about the efficacy of PBO and the involvement of monooxygenases are compounded by renewed interest in the role of esterases in pyrethroid resistance in field strains of *H. virescens*. Graves *et al.* (1991) had initially found evidence for synergism of pyrethroids with DEF, inferring the involvement of esterases and confirming earlier observations (Dowd *et al.* 1987). Martin *et al.* (1997), however, showed antagonism of the esterase synergist TPP to cypermethrin action. The larval stages of a strain of *H. virescens* originally obtained from the field in Louisiana, where control with cypermethrin and thiodicarb had failed, were examined for esterases associated with resistance to these two compounds (Goh *et al.* 1995). Esterase activity against the model substrate 1-naphthyl acetate (1-NA) was elevated in whole-body homogenates of resistant insects compared with those of susceptible insects. Increased esterase activity was attributed to three esterases, Al, Bl and Cl, which were purified and compared by means of immunoblotting techniques. The most significant of these, esterase Al, was considered to share common epitopes with the resistance-associated esterase of other insects, although its role in insecticide resistance in the tobacco budworm was not entirely clear. In a very recent study, G. Shan and J. A. Ottea (personal communication) have shown that metabolism of cypermethrin in *H. virescens* larvae occurs by both oxidative and hydrolytic pathways but that the hydrolytic route appears to be the major resistance mechanism. The production of metabolites of hydrolysis in laboratory and field strains, as well as observations that suggest that both cypermethrin and 1-NA inhibit esterases in a concentration-dependent manner, provide further evidence that esterases are the major metabolic mechanisms of resistance to pyrethroids. Moreover, inhibition experiments with PBO and paraoxon and studies with 1-NA all suggest that the monooxygenase inhibitor also inhibits esterases. Such studies clearly concur with the findings of Gunning *et al.* (1996*a*) working with *H. armigera* (see below) and cast yet further doubt on the validity of using PBO as a synergist. Urgent re-evaluation of the action and usefulness of these synergists is required.

(ii) *Metabolic resistance in* H. armigera

Australia too has been a focus of considerable debate regarding the relative roles played by esterase- and cytochrome P450-mediated pyrethroid metabolism. In 1983, with the onset of resistance to synthetic pyrethroids in *H. armigera* in Australia (Gunning *et al.* 1984), three mechanisms of resistance were thought to be involved. Both a strong nerve insensitivity (*super-kdr*) and a penetration resistance (*Pen*) were believed to be present, together with a third factor overcome by PBO (*Pbo*) (Gunning *et al.* 1991). Between 1987 and 1990 these insects were again examined to determine which mechanisms were present. Both the *Pen* and *Pbo* mechanisms had increased in importance, although they conferred only a low-order resistance of around 20-fold (Gunning *et al.* 1991). The ability of PBO to completely suppress resistance to pyrethroids in strains of the insects homozygous for a metabolic detoxication mechanism was presumed to be evidence of the involvement of P450-mediated metabolic resistance (Forrester *et al.* 1993). Moreover, the relative metabolic resistance-suppressing activity of a range of 65 synergists including TCPB provided further strong evidence of the involvement of P450-mediated metabolism. Field populations of the insects were regularly tested with a discriminating dose of fenvalerate both with and without PBO. The evidence from this monitoring suggested that the PBO-suppressible resistance component was always predominant. As discussed earlier, the non-synergizable component was assumed to represent other mechanisms, in particular nerve insensitivity. Further convincing evidence that the great majority of the resistance to pyrethroids seen Australian *H. armigera* was due to enhanced oxidation came from an important examination of the structure–activity relations of a large range of pyrethroid analogues with varying acid and alcohol structures and a range of substitutions. Alterations in the alcohol moiety of the pyrethroid structure could overcome most, if not all, resistance. The nature of these changes in countering resistance provided strong evidence that the resistance was due to oxidative metabolism. All of these findings would seem to provide overwhelming, if somewhat indirect, evidence that resistance to pyrethroids in *H. armigera* in Australia was based on enhanced P450-mediated metabolism. Similar evidence has been put forward for

enhanced monooxygenase activity being a mechanism of resistance to pyrethroids in *H. armigera* from India (Phokela & Mehrohtra 1989; Kranthi *et al.* 1997) and China (Wu *et al.* 1997*a*).

This conventional view was challenged by Kennaugh *et al.* (1993) using a strain of *H. armigera* derived from field collections and subsequently backcrossed and selected. Although the resistant strain was 19- to 33-fold resistant to fenvalerate and this resistance could be eliminated with PBO, these authors could find no increased levels of P450 in the midguts of the resistant strain compared with those of the susceptible strain. Further, there was no evidence for increased permethrin detoxication in the resistant strain. Significantly, PBO increased the rates of metabolism in both susceptible and resistant strains. Evidence was obtained which suggested the involvement of a cytochrome P450 in the process of penetration of the insecticide through the insect cuticle. The action of PBO would thus be to inhibit a P450-dependent penetration resistance (Kennaugh *et al.* 1993). These findings were corroborated by Gunning *et al.* (1995), who examined esfenvalerate metabolism in a resistant strain of *H. armigera*, in which the resistance was suppressed by PBO and which lacked any nerve insensitivity. It was shown that esfenvalerate metabolism was only slightly enhanced in this resistant strain and that PBO did not inhibit this metabolism. The authors concluded that reduced penetration appeared to be an important mechanism of esfenvalerate resistance in this strain.

An important study was then published, which suggested that pyrethroid-resistant *H. armigera* in Australia have enhanced esterase activity that is due to increased production of enzymes (Gunning *et al.* 1996*a*). The most resistant individuals were shown to have an approximately 50-fold increase in esterase activity compared with susceptible populations. Moreover, resistant strains were shown to have additional esterases not detectable in susceptible populations and increased esterase hydrolysis of 1-NA was correlated with the esfenvalerate resistance factor. Furthermore, evidence was obtained which suggested that the esterase had a poor catalytic activity towards the pyrethroids and that esterases were also acting as insecticide-sequestering agents. It was concluded that detoxification by hydrolysis together with sequestration would give *H. armigera* the ability to detoxify significant quantities of fenvalerate, consistent with the large resistance factors involved. Together these findings imply that detoxication via monooxygenases is no longer, or was never, a significant mechanism of resistance to pyrethroids in *H. armigera*, a situation that is paralleled in *H. virescens* in the USA. To further emphasize this revised view of metabolic resistance to pyrethroids, Gunning *et al.* (1996*b*) have shown that PBO can suppress esterase-mediated metabolism. This crucial observation defies the conventional assumption that PBO uniquely suppresses metabolic resistance mediated by cytochrome P450 and is again mirrored in recent studies in the USA on *H. virescens* (J. A. Ottea, personal communication). More recently Gunning *et al.* (1997) showed that fenvalerate toxicity in *H. punctigera* was synergized by the esterase inhibitors DEF and profenofos and that the resistant insects had increased esterase activity to 1-NA.

(iii) Molecular studies on metabolic resistance in heliothines

The conflicting nature of these findings both in Australia and in the USA emphasizes that a definitive role in metabolic resistance for specific P450s or esterases is likely to come only from expression studies using genes cloned from pyrethroid-resistant strains.

The complete coding sequence and parts of the 3′ and 5′ non-coding regions of a mRNA coding for a cytochrome P450 from *H. armigera* was obtained (Wang & Hobbs 1995). The sequence is most similar to member of the CYP6 family and has been designated CYP6B2. The cDNA hybridizes to two major mRNAs, the larger of which is inducible by permethrin, although the levels of induction are generally low. These same authors have demonstrated much higher quantities of the larger mRNA in individual, pyrethroid-resistant larvae collected directly from the field; this result implies the involvement of this P450 in resistance. In a separate study, RT–PCR was used to clone P450 gene fragments from the RNA of a pyrethroid-resistant strain of *H. armigera* (Pittendrigh *et al.* 1997). By this method eight new P450 genes were isolated, seven from the *CYP4* family and one *CYP9*. One of these genes, *CYP4G8*, was twofold overexpressed in the resistant strain. Although no difference in expression was noted in resistant strains, *CYP9A3* appeared to be a homologue of the putatively resistance-associated *CYP9A1* of *H. virescens* (Rose *et al.* 1997) (see below). Further, the authors found non-detectable levels of expression of the *CYP6B2* isolated by Wang & Hobbs (1995) and reportedly overexpressed in resistant strains. In *H. virescens* Rose *et al.* (1997) isolated a P450 gene designated *CYP9A1*, the first member of family 9, from a pyrethroid-resistant strain. These studies indicate that both qualitative and quantitative strain-to-strain variations in P450 expression levels are important and that recombinant expression will be necessary in order to precisely define the substrate specificities and pyrethroid-metabolizing abilities of individual P450s. The ability to define the characteristics of the detoxification systems of resistant strains of insects would lead to a significant refinement in cross-resistance studies. On the basis of substrate specificity it should prove possible not only to objectively select insecticides between chemical groups but also to select more efficacious analogues from within groups.

10. DISCUSSION

As summarized in this review, *H. armigera*, *H. virescens*, and to some extent *H. zea*, have developed substantial and often uncontrollable levels of resistance to virtually all the neurotoxic insecticides that have been directed against them. Ecological and physiological aspects of the biology of these insects have made possible the emergence of pest species, which have often proved difficult to control. Continued selection with insecticides has allowed the survival of resistant populations, which have generally proved exceedingly difficult or impossible to control.

Despite this there are examples of cropping systems in which resistance is absent or in which resistance is a minor problem. The resistance-management strategy initiated in cotton in Israel to control a range of pests has left *H. armigera* there very largely susceptible to all insecticides, despite, or more probably because of, the use of a very small number of applications (Horowitz *et al.* 1995).

In other areas such as the Ivory Coast, low or restricted use of insecticides has allowed a great many years of resistance-free pest control and only now are levels of resistance beginning to rise.

Species such as *H. punctigera*, which have, by virtue of their biology and ecology, been considered capable of escaping the development of resistance, have now been shown to do so (see, for example, Gunning *et al.* 1994, 1997). It is essential that resistance-management strategies are formulated in ways that do not enhance the resistance status of such species. In the USA, where pyrethroids have been very widely used, it is generally accepted that the key feature that has prevented development of resistance in the maize earworm, *H. zea*, is its wider range of unsprayed hosts: *H. zea* attacks maize and soyabeans whereas *H. virescens* does not attack maize and prefers cotton to soyabeans. Nevertheless, *H. zea* has recently developed significant resistance to these compounds and consideration must be given to the implications of this.

The most highly imitated resistance management strategy for heliothines is that which was set up for the control of pyrethroid- and endosulfan-resistant *H. armigera* in cotton (and other crops) in Australia (Forrester *et al.* 1993). The obvious success of this highly acclaimed scheme was that control was maintained for well over ten years with insecticides to which resistance had already developed. This was achieved largely through strict observance of restrictions in use. The gradual loss of the pyrethroids to resistance and the advent of new insecticides, *Bt* and *Bt*-transgenic cotton has allowed a relaxation of the pyrethroid use strategy and control is now based on a broad range of chemical, biological and cultural methodologies. Similar types of strategy have been initiated elsewhere with varying degrees of success, as noted above.

The development of pyrethroid resistance in heliothine species in various countries around the world continues unceasingly, even in countries with management strategies, although the rate of loss of efficacy is generally slower in controlled situations. That pyrethroid-resistant insects can be found in unsprayed refugia in Australia (Forrester *et al.* 1993), and pyrethroid-susceptible insects are absent or exceedingly hard to find anywhere in countries such as Pakistan and India, does not bode well for much further use of these compounds unless new chemistry is deployed or severe restrictions on use are instigated. Such actions have complex economic and political implications. Moreover, questions regarding the fitness costs of these resistances need to be addressed urgently since there are important considerations for insecticide-resistance management. Although new technology and new chemistry will enable the selection pressure from older insecticides to be relaxed, it is likely that the use of conventional insecticide chemistry will continue for many years.

This review has placed considerable emphasis on the mechanisms of resistance to the insecticide groups considered. This is because patterns of cross-resistance within and between insecticide groups are entirely dependent on the biochemical and molecular nature of the resistance mechanism. Even within the conventional insecticide groups there are many areas of susceptibility to existing chemistry, which can be exploited for control. In a previous brief review this author considered that target-site resistance to pyrethroids had developed in many heliothines before the later emergence of metabolic resistance (McCaffery 1994). Although this may still hold true to some degree, the ability of these species to diversify their mechanisms of resistance under selection pressure is noteworthy. The current debate over the nature of metabolic resistance in heliothines is confused by technical controversy over the use of some of the basic tools, such as synergists and model substrates, and these arguments have been considered in detail in this review. Likewise, the inconsistencies found in pyrethroid target-site mutations are at variance with those found in other insects. It is the view of this author that the heliothines are especially flexible in the use of a variety of modifications in all of their resistance mechanisms. Thus, even when a similar system of enhanced enzyme activity is involved in resistance to the same group of compounds, such as monooxygenases, different P450 forms seem likely to be found in individual populations of the same species.

It could be argued that an ability to diagnose the precise nature of the mechanisms of resistance would be a key component of the management of resistance in the heliothines. However, as emphasized in this review, the very diverse nature of the modifications found so far makes this enormously less easy than might otherwise be so and possibly renders such an approach not practicable. It might be considered that this diversity would allow the use of resistance-breaking molecules within existing conventional insecticide groups. The existence of such compounds has been illustrated for both *H. armigera* (Forrester *et al.* 1993; J. Tan and A. R. McCaffery, unpublished observations) and *H. virescens* (Shan *et al.* 1997), but the usefulness of such an approach again depends entirely on an ability to correctly diagnose subtle changes in resistance mechanisms in field populations. As highlighted above, this might prove difficult in reality and it is disappointing that such approaches have not yet resulted in commercially viable products. A knowledge of the mechanisms of resistance existing in these insects is clearly of value in devising new insecticides to control them. The advent of new areas of insecticide chemistry such as *Bacillus thuringiensis*, pyrroles, phenyl pyrazoles, spinosad, nicotinyls and insect growth regulators (IGRs) should make control of heliothines (Tabashnik *et al.*, this issue) considerably more effective and release selection on existing resistance mechanisms. Nevertheless, incipient resistance to some of these materials in a number of artificial laboratory strains of Heliothinae is surely a stimulus to effective use of new materials and an ever-watchful study of the development and nature of possible resistances to them.

In conclusion, it is perhaps significant that, to date, the most successful resistance-management programme that has been developed for these insects has been instituted in cotton in Israel. Key features of this programme have been a dramatic reduction in the number of sprays directed at a number of pests, including *H. armigera*, together with the considered use of a range of integrated pest management (IPM) techniques, It would appear that, despite our rapidly increasing knowledge of the biochemical and molecular nature of this problem, the most effective means of managing resistance to insecticides in the Heliothinae remains a strict control of insecticide use.

I am very grateful to Dr Neil Forrester, Dr James Ottea and Dr Yidong Wu for information on the current resistance situation in their countries and for permission to quote unpublished information and from papers currently in press.

REFERENCES

Abd-Elghafar, S. F., Knowles, C. O. & Wall, M. L. 1993 Pyrethroid resistance in two field strains of *Helicoverpa zea* (Lepidoptera: Noctuidae). *J. Econ. Entomol.* **86**, 1651–1655.

Abd-Elghafar, S. F., Abo-Elghar, G. E. & Knowles, C. O. 1994 Fenvalerate penetration, metabolism and excretion in pyrethroid-susceptible and resistant *Heliothis virescens* (Lepidoptera: Noctuidae). *J. Econ. Entomol.* **87**, 872–878.

Ahmad, M. & McCaffery, A. R. 1988 Resistance to insecticides in a Thailand strain of *Heliothis armigera* (Hubner) (Lepidoptera: Noctuidae). *J. Econ. Entomol.* **81**, 45–48.

Ahmad, M. & McCaffery, A. R. 1991 Elucidation of detoxication mechanisms involved in resistance to insecticides in the third instar larvae of a field-selected strain of *Helicoverpa armigera* with the use of synergists. *Pestic. Biochem. Physiol.* **41**, 41–52.

Ahmad, M., Gladwell, R. T. & McCaffery, A. R. 1989 Decreased nerve sensitivity is a mechanism of resistance in a pyrethroid resistant strain of *Heliothis armigera* from Thailand. *Pestic. Biochem. Physiol.* **35**, 165–171.

Ahmad, M., Arif, M. I. & Ahmad, Z. 1995 Monitoring insecticide resistance of *Helicoverpa armigera* (Lepidoptera: Noctuidae) in Pakistan. *J. Econ. Entomol.* **88**, 771–778.

Ahmad, M., Arif, M. I. & Attique, M. R. 1997 Pyrethroid resistance of *Helicoverpa armigera* (Lepidoptera: Noctuidae) in Pakistan. *Bull. Entomol. Res.* **87**, 343–347.

Ahmad, M., Arif, M. I., Ahmad, Z. & Attique, M. R. 1998 *Helicoverpa armigera* resistance to insecticides in Pakistan. In *Proceedings of the Beltwide Cotton Production Research Conference.* Memphis, TN: National Cotton Council. (In the press.)

Alaux, T., Vassal, J. M. & Vaissayre, M. 1997 Suivi de la sensibilité aux pyrethrinoides chez *Helicoverpa armigera* (Hubner) (Lepidoptera: Noctuidae) en Côte d'Ivoire. *J. Afric. Zool.* **111**, 63–69.

Anon 1974 Control of the resistance cotton bollworm (*Heliothis armigera* Hubner). *Kunchong Zhisi* **1**, 5–6.

Anon 1986 *Worldwide resistance to synthetic pyrethroids.* Technical Paper, Union Carbide Agricultural Products Co., USA.

Armes, N. J., Jadhav, D. R., Bond, G. S. & King, A. B. S. 1992 Insecticide resistance in *Helicoverpa armigera* in South India. *Pestic. Sci.* **34**, 355–364.

Armes, N. J., Jadhav, D. R. & De Souza, K. R. 1996 A survey of insecticide resistance in *Helicoverpa armigera* in the Indian subcontinent. *Bull. Entomol. Res.* **86**, 499–514.

Bagwell, R. D., Graves, J. B., Holloway, J. W., Leonard, B. R., Burris, E., Micinski, S. & Mascarenhas, V. 1997 Status of resistance in tobacco budworm and bollworm in Louisiana during 1996. In *Proceedings of the Beltwide Cotton Production Research Conference*, p. 1282. Memphis, TN: National Cotton Council.

Bigley, W. S. & Plapp, F. W. Jr 1978 Metabolism of *cis-* and *trans-*[14C] permethrin by the tobacco budworm and the bollworm. *J. Agric. Food Chem.* **26**, 1128–1134.

Bloomquist, J. R. 1996 Ion channels as targets for insecticides. *A. Rev. Entomol.* **41**, 163–190.

Bloomquist, J. R. & Miller, T. A. 1985 A simple bioassay for detecting and characterising insecticide resistance. *Pestic. Sci.* **16**, 611–614.

Brown, T. M. 1981 Countermeasures for insecticide resistance. *Bull. Entomol. Soc. Am.* **16**, 147–153.

Brown, T. M. & Bryson, P. K. 1992 Selective inhibitors of methyl parathion-resistant acetylcholinesterase from *Heliothis virescens. Pestic. Biochem. Physiol.* **44**, 155–164.

Brown, T. M., Bryson, P. K., Arnette, F., Roof, M., Mallett, J. L. B., Graves, J. B. & Nemec, S. J. 1996*a* Surveillance of resistant acetylcholinesterase in *Heliothis virescens*. In *Molecular genetics and evolution of pesticide resistance*, vol. 645 (ed. T. M. Brown), pp. 149–159. Washington, DC: American Chemical Society.

Brown, T. M., Bryson, P. K. & Payne, G. T. 1996*b* Synergism by propynyl aryl ethers in permethrin-resistant tobacco budworm larvae, *Heliothis virescens*. *Pestic. Sci.* **43**, 323–331.

Bull, D. L. 1981 Factors that influence tobacco budworm resistance to organophosphorus insecticides. *Bull. Entomol. Soc. Am.* **27**, 193–197.

Cameron, P. J., Walker, G. P. & Herman, T. J. B. 1995 Development of resistance to fenvalerate in *Helicoverpa armigera* (Lepidoptera: Noctuidae) in new Zealand. *NZ J. Crop Hortic. Sci.* **23**, 429–436.

Church, C. J. & Knowles, C. O. 1992 Saxitoxin binding to neural membranes from pyrethroid susceptible and resistant tobacco budworm moths *Heliothis virescens*. *Comp. Biochem. Physiol.* C **103**, 495–498.

Church, C. J. & Knowles, C. O. 1993 Relationship between pyrethroid enhanced batrachotoxin A 20-α-benzoate binding and pyrethroid toxicity to susceptible and resistant tobacco budworm moths *Heliothis virescens*. *Comp. Biochem. Physiol.* C **104**, 279–287.

Clarke, S. E., Walker, C. H. & McCaffery, A. R. 1990 A comparison of the in vitro metabolism of cis-cypermethrin in a resistant and susceptible strain of *Heliothis virescens*. In *Proc. Brighton Crop Protect. Conference: Pests and Diseases*, pp. 1201–1206. Farnham, UK: The British Crop Protection Council.

Clower, D. F., Rogers, B., Mullins, W., Marsden, D., Staetz, C. A., Monke, B. J., Phelps, J. & Certain, G. 1992 Status of *Heliothis*/*Helicoverpa* resistance to pyrethroids in US cotton: PEG-US 1991 update. In *Proceedings of the Beltwide Cotton Production Research Conference*, pp. 739–742. Memphis, TN: National Cotton Council.

Dhingra, S., Phokela, A. & Mehrohtra, K. N. 1988 Cypermethrin resistance in the populations of *Heliothis armigera* Hubner. *Natn. Acad. Sci. Lett.* **11**, 123–125.

Dowd, P. F., Gagne, C. C. & Sparks, T. C. 1987 Enhanced pyrethroid hydrolysis in pyrethroid-resistant larvae of the tobacco budworm, *Heliothis virescens* (F.). *Pestic. Biochem. Physiol.* **28**, 9–16.

Doyle, K. & Knipple, D. C. 1991 PCR-based phylogentic walking: isolation of para-homologous sodium channel gene sequences from seven insect species and an arachnid. *Insect Biochem.* **21**, 689–696.

Elliott, M. 1989 The pyrethroids: early discovery, recent advances and the future. *Pestic. Sci.* **27**, 337–351.

Elzen, G. W., Leonard, B. R., Graves, J. B., Burris, E. & Micinski, E. 1992 Resistance to pyrethroid, carbamate, and organophosphate insecticides in field populations of tobacco budworm (Lepidoptera: Noctuidae) in 1990. *J. Econ. Entomol.* **85**, 2064–2072.

Elzen, G. W., Martin, S. H. & Graves, J. B. 1993 Characteristics of tobacco budworm resistance: seasonal aspects and synergism. In *Proceedings of the Beltwide Cotton Production Research Conference*, pp. 1024–1028. Memphis, TN: National Cotton Council.

Ernst, G. & Dittrich, V. 1992 Comparative measurements of resistance to insecticides in three closely related Old and New World bollworm species. *Pestic. Sci.* **34**, 147–152.

Fitt, G. P. 1989 The ecology of *Heliothis* species in relation to agroecosystems. *A. Rev. Entomol.* **34**, 17–52.

Forrester, N. W. 1990 Designing, implementing and servicing an insecticide resistance management strategy. *Pestic. Sci.* **28**, 167–179.

Forrester, N. W., Cahill, M., Bird, L. J. & Layland, J. K. 1993 Management of pyrethroid and endosulfan resistance in *Helicoverpa armigera* (Lepidoptera: Noctuidae) in Australia. *Bull. Entomol. Res.* (Suppl.) **1**, 1–132.

Gilbert, R., Bryson, D., Bryson, P. K. & Brown, T. M. 1996 Linkage of acetylcholinesterase insensitivity to methyl parathion in *Heliothis virescens*. *Biochem. Genet.* **34**, 297–312.

Gladwell, R. T., McCaffery, A. R. & Walker, C. H. 1990 Nerve insensitivity to cypermethrin in field and laboratory strains of *Heliothis virescens*. In *Proceedings of the Beltwide Cotton Production Research Conference*, pp. 173–177. Memphis, TN: National Cotton Council.

Glenn, D. C., Hoffman, A. A. & McDonald, G. 1994 Resistance to pyrethroids in *Helicoverpa armigera* (Lepidoptera: Noctuidae) from corn: adult resistance, larval resistance and fitness effects. *J. Econ. Entomol.* **87**, 1165–1171.

Goh, D. K. S., Anspaugh, D. D., Motoyama, N., Rock, G. C. & Roe, R. M. 1995 Isolation and characterisation of an insecticide-resistance associated esterase in the tobacco budworm *Heliothis virescens* (F.). *Pestic. Biochem. Physiol.* **51**, 192–204.

Goodyer, C. J. & Greenup, L. R. 1980 A survey of insecticide resistance in the cotton bollworm, *Heliothis armigera* (Hubner) (Lepidoptera: Noctuidae) in New South Wales. *Gen. Appl. Entomol.* **12**, 37–39.

Goodyer, C. J., Wilson, A. G. L., Attia, F. I. & Clift, A. D. 1975 Insecticide resistance in *Heliothis armigera* (Hubner) (Lepidoptera: Noctuidae) in the Namoi Valley of New South Wales, Australia. *J. Aust. Entomol. Soc.* **14**, 171–173.

Gould, F. & Hodgson, E. 1980 Mixed function oxidase and glutathione transferase activity in last instar *Heliothis virescens* larvae. *Pestic. Biochem. Physiol.* **13**, 34–40.

Graves, J. B., Roussel, J. S. & Phillips, J. R. 1963 Resistance to some chlorinated hydrocarbon insecticides in the bollworm, *Heliothis zea*. *J. Econ. Entomol.* **56**, 442–444.

Graves, J. B., Leonard, B. R., Micinski, S., Long, D. & Burris, E. 1991 Status of pyrethroid resistance in tobacco budworm and bollworm in Louisiana. In *Proceedings of the Beltwide Cotton Production Research Conference*, pp. 638–641. Memphis, TN: National Cotton Council.

Graves, J. B., Leonard, B. R., Micinski, S., Burris, E., Martin, S. H., White, C. A. & Baldwin, J. L. 1993 Monitoring insecticide resistance in tobacco budworm and bollworm in Louisiana. In *Proceedings of the Beltwide Cotton Production Research Conference*, pp. 788–794. Memphis, TN: National Cotton Council.

Graves, J. B., Leonard, B. R., Burris, E., Micinski, S., Martin, S. H., White, C. A. & Baldwin, J. L. 1994 Status of insecticide resistance in tobacco budworm and bollworm in Louisiana. In *Proceedings of the Beltwide Cotton Production Research Conference*, pp. 769–774. Memphis, TN: National Cotton Council.

Gunning, R. V. 1996 Bioassay for detecting pyrethroid nerve insensitivity in Australian *Helicoverpa armigera* (Lepidoptera: Noctuidae). *J. Econ. Entomol.* **89**, 816–819.

Gunning, R. V. & Easton, C. S. 1989 Pyrethroid resistance in *Heliothis armigera* (Hubner) collected from unsprayed maize crops in New South Wales 1983–1987. *J. Aust. Entomol. Soc.* **28**, 57–61.

Gunning, R. V. & Easton, C. S. 1993 Resistance to organophosphate insecticides in *Helicoverpa armigera* (Hubner) (Lepidoptera: Noctuidae) in Australia. *Gen. Appl. Entomol.* **25**, 27–34.

Gunning, R. V. & Easton, C. S. 1994 Response of *Helicoverpa punctigera* (Wallengren) (Lepidoptera: Noctuidae) to pyrethroids, DDT and endosulfan. *J. Aust. Entomol. Soc.* **33**, 9–12.

Gunning, R. V., Easton, C. S. Greenup, L. R. & Edge, V. E. 1984 Pyrethroid resistance in *Heliothis armiger* (Hubner) (Lepidoptera: Noctuidae) in Australia. *J. Econ. Entomol.* **77**, 1283–1287.

Gunning, R. V., Easton, C. S., Balfe, M. E. & Ferris, I. G. 1991 Pyrethroid resistance mechanisms in Australian *Helicoverpa armigera*. *Pestic. Sci.* **33**, 473–490.

Gunning, R. V., Balfe, M. E. & Easton, C. S. 1992 Carbamate resistance in *Helicoverpa armigera* (Hubner) (Lepidoptera: Noctuidae) in Australia. *J. Aust. Entomol. Soc.* **31**, 97.

Gunning, R. V., Ferris, I. G. & Easton, C. S. 1994 Toxicity, penetration, tissue distribution and metabolism of methyl parathion in *Helicoverpa armigera* and *H. Punctigera* (Lepidoptera: Noctuidae). *J. Econ. Entomol.* **87**, 1180–1184.

Gunning, R. V., Devonshire, A. L. & Moores, G. D. 1995 Metabolism of esfenvalerate by pyrethroid-susceptible and -resistant Australian *Helicoverpa armigera* (Lepidoptera: Noctuidae). *Pestic. Biochem. Physiol.* **51**, 205–213.

Gunning, R. V., Moores, G. D. & Devonshire, A. L. 1996*a* Esterases and esfenvalerate resistance in Australian *Helicoverpa armigera* (Hubner) (Lepidoptera: Noctuidae). *Pestic. Biochem. Physiol.* **54**, 12–23.

Gunning, R. V., Moores, G. D. & Devonshire, A. L. 1996*b* Insensitive acetylcholinesterase and resistance to thiodicarb in Australian *Helicoverpa armigera* Hubner (Lepidoptera: Noctuidae). *Pestic. Biochem. Physiol.* **55**, 21–28.

Gunning, R. V., Moores, G. D. & Devonshire, A. L. 1997 Esterases and fenvalerate resistance in a field population of *Helicoverpa punctigera* (Lepidoptera: Noctuidae) in Australia. *Pestic. Biochem. Physiol.* **58**, 155–162.

Hardwick, D. F. 1965 The corn earworm complex. *Mem. Entomol. Soc. Canada* **40**, 1–247.

Harold, J. A. & Ottea, J. A. 1997 Toxicological significance of enzyme activities in profenofos-resistant tobacco budworms, *Heliothis virescens* (F.). *Pestic. Biochem. Physiol.* **58**, 23–33.

Harris, F. A. 1972 Resistance to methyl parathion and toxaphene-DDT in bollworm and tobacco budworm from cotton in Mississippi. *J. Econ. Entomol.* **65**, 1193–1194.

Head, D. J. 1998 *Molecular basis of nerve insensitivity resistance to pyrethroid insecticides in* Heliothis virescens *(Fabricius) and* Helicoverpa armigera *(Hubner) (Lepidoptera: Noctuidae).* PhD thesis, University of Reading.

Head, D. J., McCaffery, A. R. & Callaghan, A. 1998 Novel mutations in the *para*-homologous sodium channel gene associated with phenotypic expression of nerve insensitivity resistance to pyrethroids in heliothine Lepidoptera. *Insect Molec. Biol.* **7**, 191–196.

Holloway, J. W. & McCaffery, A. R. 1996 Nerve insensitivity to *cis*-cypermethrin is expressed in adult *Heliothis virescens*. *Pestic. Sci.* **47**, 205–211.

Holloway, J. W., Ottea, J. A. & Leonard, B. R. 1997 Mechanisms of resistance to pyrethroids in the bollworm *Helicoverpa zea*. In *Proceedings of the Beltwide Cotton Production Research Conference*, pp. 1007–1010. Memphis, TN: National Cotton Council.

Horowitz, A. R., Seligman, I. M., Forer, G., Bar, D. & Ishaaya, I. 1993 Preventative insecticide resistance management strategy in *Helicoverpa (Heliothis) armigera* (Lepidoptera: Noctuidae) in Israeli cotton. *J. Econ. Entomol.* **86**, 205–212.

Horowitz, A. R., Forer, G. & Ishaaya, I. 1995 Insecticide resistance management as a part of an IRM strategy in Israeli cotton fields. In *Challenging the future: Proceedings of the World Cotton Research Conference*, vol. 1 (ed. G. A. Constable & N. W. Forrester), pp. 537–544. Australia: CSIRO.

Ibrahim, S. A. & Ottea, J. A. 1995 Biochemical and toxicological studies with laboratory and field populations of *Heliothis virescens* (F.). *Pestic. Biochem. Physiol.* **53**, 116–128.

Jadhav, D. R. & Armes, N. J. 1996 Comparative status of insecticide resistance in the *Helicoverpa* and *Heliothis* species (Lepidoptera: Noctuidae) of south India. *Bull. Entomol. Res.* **86**, 525–531.

Kanga, L. H. B. & Plapp, F. W. Jr 1992 Development of a glass vial technique for monitoring resistance to organophosphate and carbamate insecticides in the tobacco budworm and the boll weevil. In *Proceedings of the Beltwide Cotton Production Research Conference*, pp. 731–734. Memphis, TN: National Cotton Council.

Kanga, L. H. B. & Plapp, F. W. Jr 1995 Target-site insensitivity as the mechanism of resistance to organophosphorus, carbamate and cyclodiene insecticides in tobacco budworm adults. *J. Econ. Entomol.* **88**, 1150–1157.

Kanga, L. H. B., Plapp, F. W. Jr, Elzen, G. W., Wall, M. L. & Lopez, J. D. 1995 Monitoring for resistance to organophosphorus, carbamate and cyclodiene insecticides in tobacco budworm adults (Lepidoptera: Noctuidae). *J. Econ. Entomol.* **88**, 198–204.

Kanga, L. H. B., Plapp, F. W. Jr, McCutcheon, B. F., Bagwell, R. D. & Lopez, J. D. Jr 1996 Tolerance to cypermethrin and endosulfan in field populations of the bollworm (Lepidoptera: Noctuidae) from Texas. *J. Econ. Entomol.* **89**, 583–589.

Kay, I. R. 1977 Insecticide resistance in *Heliothis armigera* (Hubner) (Lepidoptera: Noctuidae) in areas of Queensland, Australia. *J. Aust. Entomol. Soc.* **16**, 43–45.

Kay, I. R., Greenup, L. R. & Easton, C. 1983 Monitoring *Heliothis armigera* (Hubner) strains from Queensland for insecticide resistance. *Qld J. Agric. Anim. Sci.* **40**, 23–26.

Kennaugh, L., Pearce, D., Daly, J. C. & Hobbs, A. A. 1993 A piperonyl butoxide synergizable resistance to permethrin in *Helicoverpa armigera* which is not due to increased detoxification by cytochrome P450. *Pestic. Biochem. Physiol.* **45**, 234–241.

King, A. B. S. 1994 *Heliothis/Helicoverpa* (Lepidoptera: Noctuidae). In *Insect pests of cotton* (ed. G. A. Matthews & J. P. Tunstall), pp. 39–106. Wallingford, UK: CAB International.

Kirby, M. L., Young, R. J. & Ottea, J. A. 1994 Mixed-function oxidase and glutathione S-transferase activities from field-collected larval and adult tobacco budworms, *Heliothis virescens* (F.). *Pestic. Biochem. Physiol.* **49**, 24–36.

Konno, T., Hodgson, E. & Dauterman, W. C. 1989 Studies on methyl parathion resistance in *Heliothis virescens*. *Pestic. Biochem. Physiol.* **33**, 189–199.

Kranthi, K. R., Armes, N. J., Rao, N. G. V., Raj, S. & Sundaramurthy, V. T. 1997 Seasonal dynamics of metabolic mechanisms mediating pyrethroid resistance in *Helicoverpa armigera* in central India. *Pestic. Sci.* **50**, 91–98.

Lee, K. S., Walker, C. H., McCaffery, A. R., Ahmad, M. & Little, E. J. 1989 Metabolism of *trans*-cypermethrin by *Heliothis armigera* and *H. virescens*. *Pestic. Biochem. Physiol.* **34**, 49–57.

Leonard, B. R., Graves, J. B., Sparks, T. C. & Pavloff, A. M. 1988 Evaluation of field populations of tobacco budworm and bollworm (Lepidoptera: Noctuidae) for resistance to selected insecticides. *J. Econ. Entomol.* **81**, 1521–1528.

Leonova, I. N. & Slynko, N. M. 1996 Comparative study of insecticide susceptibility and activities of detoxification enzymes in larvae and adults of cotton bollworm *Heliothis armigera*. *Arch. Insect Biochem. Physiol.* **32**, 157–172.

Little, E. J., McCaffery, A. R., Walker, C. H. & Parker, T. 1989 Evidence for an enhanced metabolism of cypermethrin by a monooxygenase in a pyrethroid-resistant strain of the tobacco budworm (*Heliothis virescens* F.). *Pestic. Biochem. Physiol.* **34**, 58–68.

Luttrell, R. G., Roush, R. T., Ali, A., Mink, J. S., Reid, M. R. & Snodgrass, G. L. 1987 Pyrethroid resistance in field populations of *Heliothis virescens* (Lepidoptera: Noctuidae) in Mississippi in 1986. *J. Econ. Entomol.* **80**, 985–989.

Martin, S. H., Elzen, G. W., Graves, J. B., Micinski, S., Leonard, B. R. & Burris, E. 1992 Toxicological responses of tobacco budworms from Louisiana, Mississippi and Texas to selected insecticides. In *Proceedings of the Beltwide Cotton Production Research Conference*, pp. 735–738. Memphis, TN: National Cotton Council.

Martin, S. H., Graves, J. B., Leonard, B. R., Burris, E., Micinski, S. & Ottea, J. A. 1994 Evaluation of insecticide resistance and the effect of several synergists in tobacco budworm. In *Proceedings of the Beltwide Cotton Production Research Conference*, pp. 818–823. Memphis, TN: National Cotton Council.

Martin, S. H., Elzen, G. W., Graves, J. B., Micinski, S., Leonard, B. R. & Burris, E. 1995 Toxicological responses of tobacco budworm (Lepidoptera: Noctuidae) from Louisiana, Mississippi and Texas to selected insecticides. *J. Econ. Entomol.* **88**, 505–511.

Martin, S. H., Ottea, J. A., Leonard, B. R., Graves, J. B., Burris, E., Micinski, S. & Church, G. E. 1997 Effects of selected synergists on insecticide toxicity in tobacco budworm

(Lepidoptera: Noctuidae) in laboratory and field studies. *J. Econ. Entomol.* **90**, 723–731.

Martin, T. & Renou, A. 1995 Evolution of tolerance against chemical insecticides in two species of cotton bollworms in Chad. *Med. Fac. Landbou. Toeg. Biol. Weten Univ. Gent* **60**, 953–959.

Martinez-Carrillo, J. L. 1991 Monitoring pyrethroid resistance in the tobacco budworm *Heliothis virescens* Lepidoptera: Noctuidae in northwestern Mexico. *Southwest. Entomol. Suppl.* **15**, 59–67.

Martinez-Carrillo, J. L. 1995 Status of tobacco budworm pyrethroid resistance in Mexico. In *Challenging the future: Proceedings of the World Cotton Research Conference*, vol. 1 (ed. G. A. Constable & N. W. Forrester), pp. 545–549. Australia: CSIRO.

Martinez-Carrillo, J. L. & Reynolds, H. T. 1983 Dosage mortality studies with pyrethroids and other insecticides on the tobacco budworm (Lepidoptera: Noctuidae) from the Imperial Valley, California. *J. Econ. Entomol.* **76**, 983–986.

Martinez-Torres, D., Devonshire, A. L. & Williamson, M. S. 1997 Molecular studies of knockdown resistance to pyrethroids: cloning of domain II sodium channel gene sequences from insects. *Pestic. Sci.* **51**, 265–270.

McCaffery, A. R. 1994 Mechanisms of resistance to pyrethroids in *Helicoverpa* and *Heliothis* species. In *Proceedings of the Beltwide Cotton Production Research Conference*, pp. 836–837. Memphis, TN: National Cotton Council.

McCaffery, A. R. & Holloway, J. W. 1992 Identification of mechanisms of resistance in larvae of the tobacco budworm *Heliothis virescens* from cotton field populations. *Proc. Brighton Crop Protect. Conf.: Pests and Diseases*, pp. 227–232. Farnham, UK: The British Crop Protection Council.

McCaffery, A. R., Maruf, G. M., Walker, A. J. & Styles, K. 1988 Resistance to pyrethroids in *Heliothis* spp.: bioassay methods and incidence in populations from India and Asia. In *Proc. Brighton Crop Protect. Conference: Pests and Diseases*, pp. 433–438. Farnham, UK: The British Crop Protection Council.

McCaffery, A. R., King, A. B. S., Walker, A. J. & El-Nayir, H. 1989 Resistance to synthetic pyrethroids in the bollworm *Heliothis armigera* from Andhra Pradesh, India. *Pestic. Sci.* **27**, 65–76.

McCaffery, A. R., Walker, A. J. & Topper, C. P. 1991*a* Insecticide resistance in the bollworm *Heliothis armigera* from Indonesia. *Pestic. Sci.* **31**, 41–52.

McCaffery, A. R., Gladwell, R. T., El-Nayir, H., Walker, C. H., Perry, J. N. & Miles, M. M. 1991*b* Mechanisms of resistance to pyrethroids in laboratory and field strains of *Heliothis virescens*. *Southwest. Entomol. Suppl.* **15**, 143–158.

McCaffery, A. R., Walker, C. H., Clarke, S. E. & Lee, K. S. 1991*c* Enzymes and resistance to insecticides in *Heliothis virescens*. *Biochem. Soc. Trans.* **19**, 762–767.

McCaffery, A. R., Holloway, J. W. & Gladwell, R. T. 1995 Nerve insensitivity resistance to cypermethrin in larvae of the tobacco budworm *Heliothis virescens* from USA cotton field populations. *Pestic. Sci.* **44**, 237–247.

McCaffery, A. R., Head, D., Tan, J., Dubbeldam, A. A., Subramaniam, V. R. & Callaghan, A. 1997 Nerve insensitivity resistance to pyrethroids in heliothine Lepidoptera. *Pestic. Sci.* **51**, 315–320.

Mehrohtra, K. N. & Phokela, A. 1992 Pyrethroid resistance in *Helicoverpa armigera* Hubner. V. Response of populations in Punjab in cotton. *Pestic. Res. J.* **4**, 59–61.

Mitter, C., Poole, R. W. & Matthews, M. 1993 Biosystematics of the Heliothinae (Lepidoptera: Noctuidae). *A. Rev. Entomol.* **38**, 207–225.

Morton, N. & Collins, M. D. 1989 Managing the pyrethroid revolution in cotton. In *Pest management in cotton* (ed. M. B. Green & D. J. de B. Lyon), pp. 153–165. UK: Ellis Horwood.

Mullins, J. W., Riley, S. L., Staetz, C. A., Marrese, R. J., Rogers, B. & Monke, B. J. 1991 Status of *Heliothis* resistance to

pyrethroids in US cotton: a report from PEG-US. In *Proceedings of the Beltwide Cotton Production Research Conference*. Memphis, TN: National Cotton Council.

Narahashi, T. 1992 Nerve membrane Na$^+$ channels as targets of insecticides. *Trends Pharmacol. Sci.* **13**, 236–241.

Nicholson, R. A. & Miller, T. A. 1985 Multifactorial resistance to *trans*permethrin in field-collected strains of the tobacco budworm *Heliothis virescens* F. *Pestic. Sci.* **16**, 561–570.

Ottea, J. A., Younis, A. M., Ibrahim, S. A., Young, R. J., Leonard, B. R. & McCaffery, A. R. 1995 Biochemical and physiological mechanisms of pyrethroid resistance in *Heliothis virescens* (F.). *Pestic. Biochem. Physiol.* **51**, 117–128.

Park, Y. & Taylor, M. F. J. 1997 A novel mutation L1029H in sodium channel gene *hscp* associated with pyrethroid resistance for *Heliothis virescens* (Lepidoptera: Noctuidae). *Insect Biochem. Molec. Biol.* **27**, 9–13.

Park, Y., Taylor, M. F. J. & Feyereisen, R. 1997 A valine421 to methionine mutation in IS6 of the *hscp* voltage-gated sodium channel associated with pyrethroid resistance in *Heliothis virescens* F. *Biochem. Biophys. Res. Commun.* **239**, 688–691.

Payne, G. T. & Brown, T. M. 1984 EPN and S,S,S-tributyl phosphorotrithioate as synergists of methyl parathion in resistant tobacco budworm larvae (Lepidoptera: Noctuidae). *J. Econ. Entomol.* **7**, 294–297.

Phokela, A. & Mehrohtra, K. N. 1989 Pyrethroid resistance in *Heliothis armigera* Hubner. II. Permeability and metabolism of cypermethrin. *Proc. Natn. Acad Sci. India* B **55**, 235–238.

Phokela, A., Dhingra, S. & Mehrohtra, K. N. 1989 Pyrethroid resistance in *Heliothis armigera* Hubner. I. Response to cypermethrin. *Proc. Natn. Acad. Sci. India* B **59**, 373–380.

Phokela, A., Dhingra, S., Sinha, S. N. & Mehrohtra, K. N. 1990 Pyrethroid resistance in *Heliothis armigera* Hubner. III. Development of resistance in field. *Pestic. Res. J.* **2**, 28–30.

Pittendrigh, B., Aronstein, K., Zinkovsky, E., Andreev, O., Campbell, B., Daly, J., Trowell, S. & ffrench-Constant, R. H. 1997 Cytochrome P450 genes from *Helicoverpa armigera*: expression in a pyrethroid-susceptible and -resistant strain. *Insect Biochem. Molec. Biol.* **27**, 507–512.

Plapp, F. W. Jr & Campanhola, C. 1986 Synergism of pyrethroids by chlordimeform against susceptible and resistant *Heliothis*. In *Proceedings of the Beltwide Cotton Production Research Conference*, pp. 167–169. Memphis, TN: National Cotton Council.

Plapp, F. W. Jr, McWhorter, G. M. & Vance, W. H. 1987 Monitoring for pyrethroid resistance in the tobacco budworm. In *Proceedings of the Beltwide Cotton Production Research Conference*, pp. 324–326. Memphis, TN: National Cotton Council.

Plapp, F. W. Jr, Jackson, J. A., Campanhola, C., Frisbee, R. E., Graves, J. B., Luttrell, R. G., Kitten, W. F. & Wall, M. 1990 Monitoring and management of pyrethroid resistance in the tobacco budworm (Lepidoptera: Noctuidae) in Texas, Mississippi, Louisiana, Arkansas and Oklahoma. *J. Econ. Entomol.* **83**, 335–341.

Reed, W. T. 1974 *Heliothis* larvae: variation in mixed function oxidase activity as related to insecticide tolerance. *J. Econ. Entomol.* **67**, 150–152.

Rose, R. L., Barbhaiya, L., Roe, R. M., Rock, G. C. & Hodgson, E. 1995 Cytochrome P450-associated insecticide resistance and the development of biochemical diagnostic assays in *Heliothis virescens*. *Pestic. Biochem. Physiol.* **51**, 178–191.

Rose, R. L., Goh, D., Thompson, D. M., Verma, K. D., Heckel, D. G., Gahan, L. J., Roe, R. M. & Hodgson, E. 1997 Cytochrome P450 (CYP)9A1 in *Heliothis virescens*: the first member of a new CYP family. *Insect Biochem. Molec. Biol.* **27**, 605–615.

Roush, R. T. & Wolfenbarger, D. A. 1985 Inheritance of resistance to methomyl in the tobacco budworm (Lepidoptera: Noctuidae). *J. Econ. Entomol.* **78**, 1020–1022.

Ru, L., Wen, C., Run, C., Zhao, J. & Liu, A. 1997 The contribution and inheritance of *kdr* to fenvalerate and cyhalothrin resistance in *Helicoverpa armigera*. *Resist. Pest Mgmt* **9**, 9–10.

Sekhar, P. R., Venkataiah, M., Rao, N. V., Rao, B. R. & Roa, V. S. P. 1996 Monitoring of insecticide resistance in *Helicoverpa armigera* (Hubner) from areas receiving heavy insecticidal applications in Andhra Pradesh (India). *J. Entomol. Res.* **20**, 93–102.

Shan, G., Hammer, R. P. & Ottea, J. A. 1997 Biological activity of pyrethroid analogs in pyrethroid-susceptible and -resistant tobacco budworms, *Heliothis virescens* (F.). *J. Agric. Food Chem.* **45**, 4466–4473.

Shen, J., Tan, J., Xiao, B., Tan, F. & You, Z. 1991 Monitoring and forecasting of pyrethroids resistance of *Heliothis armigera* (Hubner) in China. *Kunchong Zhishi* **28**, 337–341.

Shen, J., Wu, Y., Tan, J. & Tan, F. 1992 Pyrethroid resistance in *Heliothis armigera* (Hubner) (Lepidoptera: Noctuidae) in China. *Resist. Pest Mgmt* **4**, 22–24.

Shen, J., Wu, Y., Tan, J., Zhou, B., Jin, C. & Tan, F. 1993 Comparison of two monitoring methods for pyrethroid resistance in cotton bollworm (Lepidoptera: Noctuidae). *Resist. Pest Mgmt* **5**, 5–7.

Soderlund, D. M. & Bloomquist, J. R. 1989 Neurotoxic action of pyrethroid insecticides. *A. Rev. Entomol.* **34**, 77–96.

Sparks, T. C. 1981 Development of insecticide resistance in *Heliothis zea* and *Heliothis virescens* in North America. *Bull. Entomol. Soc. Am.* **27**, 186–192.

Sparks, T. C., Graves, J. B. & Leonard, B. R. 1993 Insecticide resistance and the tobacco budworm: past, present and future. *Rev. Pestic. Toxicol.* **2**, 149–183.

Stadelbacher, E. A., Snodgrass, G. L. & Elzen, G. W. 1990 Resistance to cypermethrin in first generation adult bollworm and tobacco budworm (Lepidoptera: Noctuidae) populations collected as larvae on wild geranium, and the second and third larval generations. *J. Econ. Entomol.* **83**, 1207–1210.

Suckling, D. M. 1996 Status of insecticide and miticide resistance in New Zealand. In *Pesticide resistance: prevention and management* (ed. G. W. Bourdot & D. M. Suckling), pp. 49–58. New Zealand: Rotorua New Zealand Plant Protection Society.

Sukhoruchenko, G. I. 1996 Pesticide resistance of cotton plant pests in Central Asia and Azerbaijan: state of the problem in the early 90s. *Entomologicheskoe Oboozrenie* **75**, 3–15.

Tan, J., Tan, F. & You, Z. 1987 Monitoring and selection for resistance of cotton bollworm, *Heliothis armigera* (H.) to four pyrethroids. *J. Nanjing Agric. Univ.* **4**, 36–43.

Taylor, M. F. J., Heckel, D. G., Brown, T. M., Kreitman, M. E. & Black, B. 1993 Linkage of pyrethroid insecticide resistance to a sodium channel locus in the tobacco budworm. *Insect Biochem. Molec. Biol.* **23**, 763–775.

Twine, P. H. & Reynolds, H. T. 1980 Relative susceptibility and resistance of the tobacco budworm to methyl parathion and synthetic pyrethroids in southern California. *J. Econ. Entomol.* **73**, 239–242.

Vassal, J. M., Vaissayre, M. & Nartin, T. 1997 Decrease in the susceptibility of *Helicoverpa armigera* (Hubner) (Lepidoptera: Noctuidae) to pyrethroid insecticides in Cote d'Ivoire. *Resist. Pest Mgmt* **9**, 14–15.

Wang, X. & Hobbs, A. A. 1995 Isolation and sequence analysis of a cDNA clone for a pyrethroid inducible cytochrome P450 from *Helicoverpa armigera*. *Insect Biochem. Molec. Biol.* **25**, 1001–1009.

Wangboonkong, S. 1981 Chemical control of cotton pests in Thailand. *Trop. Pest Mgmt* **27**, 495–500.

West, A. J. & McCaffery, A. R. 1992 Evidence for nerve insensitivity to cypermethrin from Indian strains of *Helicoverpa armigera*. In *Proc. Brighton Crop Protect. Conference: Pests and Diseases*, pp. 233–238. Farnham, UK: The British Crop Protection Council.

Whitten, C. J. & Bull, D. L. 1974 Comparative toxicity, absorption and metabolism of chlorpyrifos and its dimethyl homologue in methyl parathion-resistant and -susceptible tobacco budworms. *Pestic. Biochem. Physiol.* **4**, 266–274.

Whitten, C. J. & Bull, D. L. 1978 Metabolism and absorption of methyl parathion by tobacco budworms resistant or susceptible to organophosphorus insecticides. *Pestic. Biochem. Physiol.* **9**, 196–202.

Wilkinson, I. J. & McCaffery, A. R. 1991 Titres of *cis*-cypermethrin in the CNS of resistant and susceptible strains of *Heliothis virescens*; pharmacokinetic modification of target site exposure. *Pestic. Sci.* **34**, 90–91.

Williamson, M. S., Martinez-Torres, D., Hick, C. A. & Devonshire, A. L. 1996 Identification of mutations in the housefly *para*-type sodium channel gene associated with knockdown resistance (*kdr*) to pyrethroid insecticides. *Molec. Gen. Genet.* **252**, 51–60.

Wilson, A. G. L. 1974 Resistance of *Heliothis armigera* to insecticides in the Ord irrigation area, north western Australia. *J. Econ. Entomol.* **67**, 256–258.

Wolfenbarger, D. A. & McGarr, R. L. 1970 Toxicity of methyl parathion, parathion and monocrotophos applied topically to populations of lepidopteran pests of cotton. *J. Econ. Entomol.* **63**, 1762–1764.

Wolfenbarger, D. A., Bodegas, P. R. & Flores, R. 1981 Development of resistance in *Heliothis* spp. in the Americas, Australia, Africa and Asia. *Bull. Entomol. Soc. Am.* **27**, 181–185.

Wu, K., Liang, G. & Guo, Y. 1997 Phoxim resistance in *Helicoverpa armigera* (Lepidoptera: Noctuidae) in China. *J. Econ. Entomol.* **90**, 868–872.

Wu, Y., Shen, J., Tan, F. & You, Z. 1995 Mechanism of fenvalerate resistance in *Helicoverpa armigera* Hubner. *J. Nanjing Agric. Univ.* **18**, 63–68.

Wu, Y., Shen, J., Chen, J., Lin, X. & Li, A. 1996 Evaluation of two resistance monitoring methods in *Helicoverpa armigera*: topical application method and leaf dipping method. *J. Plant Protect.* **5**, 3–6.

Wu, Y., Shen, J., Chen, J., Zhou, W. & Li, A. 1997*a* Determination and tissue distribution of microsomal cytochrome P450 and cytochrome *b*5 in six-instar larva of *Helicoverpa armigera*. *J. Agric. Biotech.* **5**, 297–301.

Wu, Y., Shen, J., Tan, F. & You, Z. 1997*b* Resistance monitoring of *Helicoverpa armigera* in Yanggu County of Shandong Province. *J. Nanjing Agric. Univ.* **18**, 48–53.

Zalucki, M. P. (ed.) 1991 *Heliothis: research methods and prospects*. New York: Springer.

Zhang, Y., Han, X., Zhang, W., Luo, L. & Zhou, P. 1997 An electrophysiological study on resistance to pyrethroid insecticides in *Helicoverpa armigera*. *Acta Entomol. Sinica* **40**, 113–121.

Zhao, G., Rose, R. L., Hodgson, E. & Roe, R. M. 1996 Biochemical mechanisms and diagnostic microassays for pyrethroid, carbamate and organophosphate insecticide resistance/cross-resistance in the tobacco budworm, *Heliothis virescens*. *Pestic. Biochem. Physiol.* **56**, 183–195.

Zhao, Y., Liu, A. & Ru, L. 1996 Decreased nerve sensitivity is an important pyrethroid resistance mechanism of cotton bollworm. *Acta Entomol. Sinica* **39**, 347–353.

Zhu, M., Zhang, D., Tan, F. & Shen, J. 1982 A study of the resistance of agricultural pests to insecticides. I. An investigation of *H. armigera* (Hubner). *J. Nanjing Agric. Coll.* **2**, 1–7.

Insect resistance to *Bacillus thuringiensis*: uniform or diverse?

Bruce E. Tabashnik[1]*, **Yong-Biao Liu**[1], **Thomas Malvar**[2], **David G. Heckel**[3], **Luke Masson**[4] **and Juan Ferré**[5]

[1]*Department of Entomology, University of Arizona, Tucson, AZ 85721, USA*
[2]*Monsanto Company, 700 Chesterfield Parkway North, St Louis, MO 63017, USA*
[3]*Department of Biological Sciences, Clemson University, Clemson, SC 29634, USA*
[4]*Biotechnology Research Institute, Montreal, Quebec, Canada H4P 2R2*
[5]*Departamento de Genètica, Universitat de València, 46100 Burjassot (València), Spain*

Resistance to the insecticidal proteins produced by the soil bacterium *Bacillus thuringiensis* (Bt) has been documented in more than a dozen species of insect. Nearly all of these cases have been produced primarily by selection in the laboratory, but one pest, the diamondback moth (*Plutella xylostella*), has evolved resistance in open-field populations. Insect resistance to Bt has immediate and widespread significance because of increasing reliance on Bt toxins in genetically engineered crops and conventional sprays. Furthermore, intense interest in Bt provides an opportunity to examine the extent to which evolutionary pathways to resistance vary among and within species of insect. One mode of resistance to Bt is characterized by more than 500-fold resistance to at least one CrylA toxin, recessive inheritance, little or no cross-resistance to CrylC, and reduced binding of at least one CrylA toxin. Analysis of resistance to Bt in the diamondback moth and two other species of moths suggests that although this particular mode of resistance may be the most common, it is not the only means by which insects can attain resistance to Bt.

Keywords: allelism; *Bacillus thuringiensis*; diamondback moth; evolution; genetic variation; resistance

1. INTRODUCTION

Widespread insecticide resistance and rising concerns about environmental hazards have spurred the search for alternatives to conventional insecticides. Thus, insecticides derived from the common bacterium *Bacillus thuringiensis* (Bt) are becoming increasingly important for pest management (Entwistle *et al.* 1993).

Bt is a natural pathogen of some pests. Insecticidal proteins produced by Bt are extremely toxic to certain pests, but cause little or no harm to people, wildlife, and most beneficial insects. Therefore, compared with many conventional insecticides, Bt-based insecticides pose less risk to the environment and are more compatible with biological control.

Bt has been used in sprays for more than 30 years, but recent breakthroughs in biotechnology have greatly enhanced the role of Bt in agriculture. Through genetic engineering, scientists have transferred genes encoding Bt toxins into crop plants. In effect, these transgenic plants produce their own environmentally benign insecticide.

Three transgenic crops that produce Bt toxins were grown commercially in the USA during 1997: nearly 3 Mha of Bt maize, 1 Mha of Bt cotton, and 10 kha of Bt potato (Wadman 1997; Mellon & Rissler 1998). These large plantings represent a huge increase in use of Bt, but only a tiny portion of the potential world market for Bt crops. At least 15 other Bt crops—including apple, broccoli, poplar, rice, tomato, and walnut—have been approved for field testing by the US Department of Agriculture (Mellon & Rissler 1998). Despite some problems, Bt crops have generally performed well, and in most cases, have greatly reduced the use of conventional insecticides.

Foliar sprays of Bt containing toxins, spores, and other materials were used for decades without any reports of pest resistance from the field. This led some to wonder whether insects could evolve resistance to Bt. With laboratory-selected resistance to Bt demonstrated in many pests and field-evolved resistance to Bt documented in the diamondback moth (*Plutella xylostella*), adaptation by pests is now considered the biggest threat to the long-term success of Bt.

Several recent reviews cover the general topics of evolution and management of resistance to Bt (Ferré *et al.* 1995; Gould 1998; McGaughey & Whalon 1992; Tabashnik 1994). Here we ask, 'how much does the genetic basis of resistance to Bt vary among populations of moths?' The following sections provide a brief overview of the biology of Bt and resistance to Bt, a summary of variation in the genetic basis of resistance to Bt among three widely separated populations of diamondback moth, and comparisons with other Bt-resistant strains of diamondback moth and other moths.

*Author for correspondence (brucet@ag.arizona.edu).

Finally, we consider the implications of the aforementioned results for prolonging the efficacy of Bt through resistance management.

2. OVERVIEW OF BT AND RESISTANCE TO BT

During sporulation, Bt produces crystalline inclusions composed of proteins called δ-endotoxins. Because they occur in crystals, these proteins are referred to as Cry toxins (Crickmore *et al.* 1998). To kill insects, Bt crystals must be ingested. In the alkaline insect midgut, crystals dissolve into protoxins, which are cleaved by proteases into active toxins (Gill *et al.* 1992). Toxins bind to and form pores in the brush border of midgut membranes; this makes cells swell and lyse, eventually causing death (Gill *et al.* 1992).

X-ray crystallography has revealed the three-dimensional structure of toxins Cry3A (Li *et al.* 1991), which kills beetles, and CryIAa (Grochulski *et al.* 1995), which kills moth larvae. Each toxin has three domains. Current thinking is that the α-helices of domain I are critical in pore formation and the loops at the ends of the β-sheets of domain II are essential for binding (Grochulski *et al.* 1995).

The only well-characterized mechanism of resistance to Bt is reduced binding of toxin to midgut membranes (Ferré *et al.* 1991, 1995; Van Rie *et al.* 1990). Nonetheless, this is not the only mechanism, as several cases of resistance to Bt are not associated with reduced binding of toxin (Ferré *et al.* 1995; Oppert *et al.* 1997; Tabashnik 1994).

Although Bt is sometimes mistakenly considered a singular entity, thousands of strains of Bt are housed in various collections maintained by industry, governments, and academia. Each strain has a characteristic set of toxins. Generally, toxins with related amino-acid sequences kill related insects. For example, CryI toxins kill larvae of some species of moths, whereas Cry3 toxins are lethal to certain beetles.

The number of reported DNA sequences for Bt toxin genes grew from four in 1985 to 135 in 1997 (Crickmore *et al.* 1998). This rapid growth has prompted the establishment of a new system of nomenclature and a web site to track the latest genes (Crickmore *et al.* 1998). The expansion includes gene variants that differ by less than 5% of their amino-acid sequence (e.g. six different forms of CryIAa), as well as discovery of major new types of Bt toxin genes.

If there are so many Bt toxin genes, with more being discovered every year, why not just switch to a new toxin when resistance occurs? Two factors limit the potential for the toxin-switching strategy. First, a relatively small subset of toxins kill any particular pest. For example, CryIC but not CryIA toxins are highly effective against the beet armyworm, *Spodoptera exigua* (Moar *et al.* 1995). Second, selection with one toxin or set of toxins can produce cross-resistance to others (Gould *et al.* 1992; McGaughey & Johnson 1992; Moar *et al.* 1995; Tabashnik *et al.* 1996). So, despite the abundance and diversity of Bt toxins, resistance remains a serious concern.

Laboratory selection experiments show that many pests, including targets of Bt crops, can readily evolve resistance to Bt (Tabashnik 1994). However, the diamondback moth offers the only opportunities now to study patterns of evolution of resistance to Bt in open-field populations. Damage inflicted and control costs for this cosmopolitan pest of cabbage and related vegetables exceed one billion US dollars annually (Talekar & Shelton 1993). Larvae of the diamondback moth eat the foliage of cruciferous vegetables and, if susceptible, are killed by Bt toxins. Field-evolved resistance to Bt has been documented in populations from the USA (Florida, Hawaii and New York), Central America (Costa Rica, Guatemala, Honduras and Nicaragua), and Asia (China, Japan, Malaysia, the Philippines and Thailand) (Perez & Shelton 1997; Tabashnik 1994; Wright *et al.* 1997; Zhao *et al.* 1993).

3. VARIATION IN BT RESISTANCE AMONG DIAMONDBACK MOTH POPULATIONS

How similar is the genetic basis of resistance to Bt in different populations of diamondback moth? Expectations depend on the frequency and type of resistance-conferring mutations (McKenzie & Batterham 1994). If mutations conferring resistance to Bt are exceedingly rare, a single mutational event might occur and the resulting resistance-conferring allele might subsequently spread worldwide by migration. This is the scenario proposed by Raymond *et al.* (1991) for one type of mosquito resistance to organophosphates. If resistance-conferring mutations are highly constrained (ffrench-Constant *et al.* 1993), yet somewhat more common, similar or identical mutations might arise independently in many populations. If mutations conferring resistance to Bt are neither exceedingly rare nor highly constrained, populations might differ in the loci at which resistance-conferring mutations occur. In this last scenario, variation among populations might occur in key traits such as the spectrum of resistance and cross-resistance, dominance, and the mechanism of resistance.

To test these ideas, a team of scientists from five research groups in three countries compared Bt-resistant strains of diamondback moth derived from Hawaii (NO-QA), Pennsylvania (PEN), and the Philippines (PHI). Here we provide highlights of experimental analyses of cross-resistance, dominance of resistance, genetic correlations among resistances to different toxins within each strain, interstrain complementation tests for allelism, and binding of toxins to their target sites in the midgut. Results reported previously (Tabashnik *et al.* 1997*a,b*), and summarized here, suggest that one pair of populations shares a locus at which similar or identical mutations can confer resistance to at least four Bt toxins. In contrast, resistance to Bt in a third population involves a different mutation at the shared locus. Further, each population has at least two independently segregating genes that can affect resistance to Bt.

(a) *Methods*

Three Bt-resistant strains were isolated from field populations that had been treated extensively with foliar applications of commercial formulations of the Bt subspecies *kurstaki* containing CryIA toxins and other materials (Tabashnik *et al.* 1997*a,b*). Before comparisons were done, each strain was subjected to selection with Bt in the laboratory to further reduce the frequency of susceptible individuals. Aside from binding data, which are summarized briefly, the results described here are from bioassays in which groups of third-instar larvae were exposed to cabbage-leaf discs that had been dipped in distilled-water dilutions of CryIAa, CryIAb, CryIAc, CryIC, CryIF, CryIJ, or distilled water only as a control. The susceptible LAB-P strain was included as a control in all bioassays. Strains were tested side by side to avoid potentially confounding effects of environmental differences between laboratories.

(b) *Spectrum of resistance and cross-resistance*

Responses of each resistant strain to six CryI toxins revealed a pattern that emerged in all of our tests: the NO-QA strain from Hawaii and the PEN strain from Pennsylvania were similar, but the PHI strain from the Philippines was different. Both NO-QA and PEN were extremely resistant to CryIAa, CryIAb, and CryIAc, susceptible to CryIC, and cross-resistant to CryIF and CryIJ. Like NO-QA and PEN, PHI was resistant to the CryIA toxins and susceptible to CryIC. However, unlike the other two strains, PHI showed no cross-resistance to CryIF or CryIJ.

(c) *Dominance*

Before reviewing the experimental results, we shall digress briefly here to define dominance and explain its significance for resistance management. The simplest genetic basis for resistance would be a single locus with one allele (R) for resistance and another for susceptibility (S). Although we know that resistance to Bt involves more than two alleles at one locus, alleles with major effects do exist (Heckel *et al.* 1997; Tabashnik *et al.* 1995, 1997*a,b*; Tang *et al.* 1997), and the simplest model is a reasonable starting point.

Because alleles for Bt resistance are rare initially (Gould *et al.* 1997), individuals homozygous for resistance to Bt (RR) are exceedingly rare initially, occurring at about the square of the frequency of the resistance allele (Tabashnik 1997). Thus, the response of heterozygotes (RS) to Bt determines the initial course of evolution of resistance. If Bt kills heterozygotes, the resistance is termed recessive. If heterozygotes survive exposure to Bt, the resistance is called dominant.

Several resistance-management strategies, including the popular 'refuge–high dose' strategy, work best if resistance is recessive. A refuge is an area in which a portion of the pest population is not exposed to Bt. Hence, refuges enable survival of susceptible individuals. The idea behind the refuge–high dose strategy is that the very rare homozygous resistant (RR) adults mate with the much more abundant homozygous susceptible (SS) adults emerging from the refuge, and these matings generate heterozygous (RS) offspring. If resistance is recessive (heterozygotes are killed by Bt) and other assumptions of the strategy are valid, this approach can substantially postpone the evolution of resistance.

To evaluate dominance, we crossed each resistant strain with the susceptible LAB-P strain. In all cases, we paired a single virgin male from one strain with a single virgin female from another strain. The progeny from each single-pair family were reared and tested separately from all other families.

As with the spectrum of resistance, NO-QA and PEN were alike in terms of dominance, but PHI was different. For NO-QA and PEN, resistance to CryIAa, CryIAb, CryIAc and CryIF was partly to completely recessive. PHI showed recessive inheritance of resistance to CryIAb, but its resistance to CryIAa and CryIAc was not recessive. Mortality caused by CryIAc to the 16 F_1 families from PHI × LAB-P ranged from 21 to 90% with a mean of 64%; this result suggests control by one or more semidominant mutations.

We were surprised to find evidence for dominant resistance to CryIAa in PHI. Mortality caused by CryIAa ranged from 0 to 10% (mean 4.6%) in 5 of the 16 single-pair F_1 families derived from PHI × LAB-P. These results show that PHI harboured at least one dominant mutation conferring resistance to CryIAa. In contrast, mean mortality caused by CryIAa averaged 76% in F_1 families from NO-QA × LAB-P and 85% in F_1 families from PEN × LAB-P. None of the 14 F_1 families from PEN × LAB-P had mortality less than 20%. One of the 13 F_1 families from NO-QA × LAB-P had 11% mortality and two others had lower than expected mortality. Unexpectedly low mortality in response to CryIAa in some F_1 families from NO-QA × LAB-P suggests that at least one non-recessive mutation conferring resistance to CryIAa also occurred in NO-QA.

(d) *Genetic correlations and numbers of loci influencing resistance within each strain*

To determine whether single genes in NO-QA could confer resistance to more than one toxin, we generated hybrids by crossing NO-QA × LAB-P, reared the hybrids for one or more generations, then selected them with a single toxin. We found that by selecting with CryIAa, we could generate cross-resistance to CryIAb, CryIAc, and CryIF. Analogously, selection with either CryIAb or CryIAc immediately produced cross-resistance to each of the other three toxins in this set. These data support the hypothesis that the NO-QA strain harbours a mutation conferring resistance to CryIAa, CryIAb, CryIAc, and CryIF.

Although the selection approach described above is a compelling way to test for the effects of a single locus on resistance to several toxins, we needed a more efficient approach to enable comparisons to be made among all three resistant strains. Thus, we used experiments in which the progeny from each single-pair were split so that groups of siblings from each family were tested with either CryIAa, CryIAb, CryIAc, or CryIF.

As with our earlier comparisons, results of the split-brood experiments showed that NO-QA and PEN were alike, yet PHI was different. Strong genetic correlations (mean $r = 0.80$) were evident between all six pairwise combinations of CryIAa, CryIAb, CryIAc, and CryIF for NO-QA and PEN, but not for PHI. The observed genetic correlations are consistent with the hypothesis that in NO-QA and PEN a single mutation can confer resistance to all four toxins. In contrast, resistance to CryIF in PHI was not correlated with resistance to the other toxins. Further, pairwise correlations between the CryIA toxins were either weak (CryIAa–CryIAc and CryIAb–CryIAc) or not significant (CryIAa–CryIAb) in PHI. Therefore, the evidence suggests that NO-QA and PEN have one or more multitoxin resistance mutations that are rare or absent in PHI.

The data suggest that NO-QA and PEN each harbour a multitoxin resistance gene that can confer resistance to at least four toxins, but they also show that each of these strains has at least two independently segregating resistance genes. As noted above, some hybrid F_1 families from NO-QA × LAB-P had unexpectedly low mortality in response to CryIAa. Likewise, some hybrid F_1 families from PEN × LAB-P had patterns of mortality that cannot be explained by allelic variation at a single locus. Overall, these patterns are reflected in significant toxin × family interactions in mortality in the hybrid F_1 families from NO-QA × LAB-P, PEN × LAB-P, and PHI × LAB-P. These interactions indicate that at least two independently segregating loci influence resistance within each strain.

(e) *Interstrain complementation tests for allelism*

We used interstrain complementation tests to determine whether the mutations conferring resistance in different strains were alleles at a shared locus. If resistance is recessive and controlled by the same locus in two different strains, hybrid progeny from crosses between the two strains will receive one allele for resistance from their father and another from their mother. Although these resistance alleles will not necessarily be identical, they will be at the same locus. Thus, lacking an allele for susceptibility at the resistance locus, the hybrid progeny will be resistant.

An alternative hypothesis is that resistance is controlled by locus 1 in resistant strain A and by independently segregating locus 2 in resistant strain B. If so, at locus 1, hybrid progeny from a cross between strains A and B will receive an allele for resistance from their strain A parent and an allele for susceptibility from their strain B parent. The converse will be true for locus 2. If resistance is recessive and epistasis is absent, such doubly heterozygous progeny will be susceptible. Thus, the shared-locus hypothesis predicts that hybrid progeny from interstrain crosses will be resistant and the independent-locus hypothesis predicts that they will be susceptible.

Our results support the shared-locus hypothesis. Hybrid progeny from NO-QA × PEN were resistant to CryIAa, CryIAb, CryIAc, and CryIF. Hybrid progeny from PHI × NO-QA and PHI × PEN were resistant to CryIAb. CryIAb was the only toxin for which PHI showed recessive inheritance of resistance and thus the only toxin for which the test for allelism was informative for PHI.

(f) *Conclusions from comparisons among Bt-resistant strains of diamondback moth*

The complementation data show that all three strains share a resistance locus. We call this gene 'BtR-1' of the diamondback moth. As described above, mutations at this locus in each strain confer resistance that is recessive. Biochemical assays described elsewhere show that this resistance is associated with reduced binding (Tabashnik *et al.* 1997*b*). The genetic correlation analyses show that the NO-QA and PEN strains have at least one multitoxin resistance allele at this locus that can confer resistance to CryIAa, CryIAb, CryIAc, and CryIF. Given that these

four toxins share a binding site in the diamondback moth (Ballester *et al.* 1994; Granero *et al.* 1996), the simplest interpretation is that, in NO-QA and PEN, one mutation can alter binding of all four toxins.

In contrast, the PHI strain has one or more alleles at the BtR-1 locus that confer resistance to CryIAb but not to CryIAa or CryIF. These data imply that an alternative mutation at the BtR-1 locus alters binding of CryIAb, but not other toxins. Results from the Bt-resistant SERD3 strain of diamondback moth from Malaysia also show reduced binding of CryIAb with no change in binding of CryIAa or CryIAc (Wright *et al.* 1997).

In summary, NO-QA and PEN share a major resistance locus at which at least one multitoxin resistance mutation occurs. These two strains are also similar in their spectrum of resistance and cross-resistance, dominance of resistance, and mechanism of resistance. The PHI strain has a different allele for resistance at the shared locus and a narrower spectrum of resistance, with no cross-resistance to CryIF or to CryIJ. Compared with resistance in NO-QA and PEN, resistance to CryIAb in PHI shows the same dominance and mechanism, but resistance to CryIAa and CryIAc does not.

The multitoxin resistance alleles at locus BtR-1 in NO-QA and PEN are similar but not necessarily identical. We suspect that they arose independently because of the geographical isolation between the source locations in Hawaii and Pennsylvania and the recent appearance of resistance to Bt. The PHI strain contains an allele at the BtR-1 locus that is different from that of NO-QA or PEN and thus must have arisen independently.

Thorough analyses of resistance to Bt in the Loxa A strain of diamondback moth, which was derived from a resistant field population in Florida (Tang *et al.* 1996, 1997), show similarities with the resistance to Bt in the strains from Hawaii and Pennsylvania described above. Although CryIF and CryIJ have not yet been tested against Loxa A, this strain from Florida is extremely resistant to CryIAa, CryIAb, and CryIAc, yet, like the other strains, it is susceptible to CryIC. Resistance to CryIAc in Loxa A is recessive and probably controlled by a single locus (Tang *et al.* 1997). Resistance to CryIAb in Loxa A is associated with reduced binding (Tang *et al.* 1996). We do not know whether a single locus confers resistance to several toxins in Loxa A or whether the mutation or mutations conferring resistance to CryIA toxins in Loxa A are allelic with those in NO-QA and PEN.

4. COMPARISONS WITH RESISTANCE TO BT IN OTHER MOTHS

Comparisons with the tobacco budworm (*Heliothis virescens*) and the Indianmeal moth (*Plodia interpunctella*), two species in which resistance to Bt has been studied intensively, reveal striking parallels with the diamondback moth. First, at least one strain of each species exhibits a similar type of resistance to CryIA toxins. We'll call this 'mode 1' of resistance to Bt, which is characterized by extremely high resistance (over 500-fold) to at least one CryIA toxin, recessive inheritance, little or no cross-resistance to CryIC, and reduced binding of at least one CryIA toxin. Mode 1 resistance has been reported for the NO-QA, PEN, and Loxa A strains of diamondback moth (Tabashnik *et al.* 1994, 1996, 1997*a*,*b*; Tang *et al.* 1996, 1997), the PHI strain of diamondback moth against CrylAb (Tabashnik *et al.* 1997*b*), the YHD2 strain of tobacco budworm (Gould *et al.* 1995; Heckel *et al.* 1997; Lee *et al.* 1995), and the 343R strain of Indianmeal moth (McGaughey 1985; Van Rie *et al.* 1990).

Second, at least one strain of each of these three moth species showed resistance to Bt that differs from mode 1. Resistance to CrylAa and CrylAc in the PHI strain of diamondback moth was not recessive and not associated with reduced binding (Tabashnik *et al.* 1997*b*). Resistance in the SEL and CP73 strains of tobacco budworm was less than 100-fold, not recessive, and not associated with reduced binding (Gould *et al.* 1992; MacIntosh *et al.* 1991; Sims & Stone 1991). Oppert *et al.* (1997) reported evidence for altered protease activity as a mechanism of resistance in Indianmeal moth strains 133-r and 198-r. These two strains, which had been selected with Bt subspecies *aizawai* and *entomocidus*, respectively, both showed resistance to CrylB and CrylC as well as to CrylA toxins (McGaughey & Johnson 1992, 1994). Other examples of resistance also show traits different from mode 1 resistance, such as resistance to CrylC in the diamondback moth (Liu & Tabashnik 1997), *Spodoptera exigua* (Moar *et al.* 1995), and *Spodoptera littoralis* (Chaufaux *et al.* 1997).

5. IMPLICATIONS FOR RESISTANCE MANAGEMENT

Because a particular pest can have more than one mode of resistance, characterization of one or a few resistant strains may not be sufficient for understanding resistance to Bt in a species. Therefore, in designing resistance-management strategies, it is unwise to assume that populations of a given pest can evolve only one type of resistance to Bt. For example, despite the finding of many cases of recessive resistance to Bt toxins in the diamondback moth, dominant resistance to CrylAa in the PHI strain violates one of the key assumptions of the refuge–high dose strategy.

In the light of this variation in resistance to Bt within species, how should we proceed? As resistance to Bt begins to evolve in the field in pests other than the diamondback moth, it will be critical to determine if mode 1 resistance is predominant. If so, it might be reasonable to implement strategies that are likely to be especially effective against this type of resistance. If diverse modes are found in other pests, as in the diamondback moth, or if a different type of resistance is most common, rethinking of strategies may be needed. In the meantime, increasing the spatial and temporal refuges from exposure to Bt should delay most, if not all, types of resistance.

In any case, admission of ignorance is important because current efforts at resistance management rely heavily on computer simulations and laboratory experiments, with few or no rigorous tests of tactics in the field. Careful tracking of resistance episodes in the field, either in conjunction with either designed experiments or the natural experiments that are proceeding on millions of hectares, can help to test predictions about evolution of resistance. Such tests are essential for enhancing the credibility of resistance management and sustaining the efficacy of benign insecticides such as Bt toxins.

We thank N. Finson, D. Coyle, K. Johnson, V. Ballester, F. Granero and J. L. Ménsua, for their invaluable contributions to the research described herein. T. Dennehy, P. Follett, B. Oppert and Y. Park provided valuable comments and suggestions. Mycogen and Ruud de Maagd generously provided toxins. The research was supported by the United States Department of Agriculture (TSTAR 95-34135-1771, NRI-CGP 96-35302-3470, and WRPIAP 97RA0304/0305-WR96-16) and the European Community (ECLAIR project AGRE-0003).

REFERENCES

Ballester, V., Escriche, B., Ménsua, J. L., Riethmacher, G. W. & Ferré, J. 1994 Lack of cross-resistance to other *Bacillus thuringiensis* crystal proteins in a population of *Plutella xylostella* highly resistant to CryIA(b). *Biocontrol Sci. Technol.* **4**, 437–443.

Chaufaux, J., Müller-Cohn, J., Buisson, C., Sanchis, V., Lereclus, D. & Pasteur, N. 1997 Inheritance of resistance to the *Bacillus thuringiensis* CrylC toxin in *Spodoptera littoralis* (Lepidoptera: Noctuidae). *J. Econ. Entomol.* **90**, 873–878.

Crickmore, N., Zeigler, D. R., Feitelson, J., Schnepf, E., Lereclus, D., Baum, J., Van Rie, J. & Dean, D. H. 1998 WWW site: http://www.biols.susx.ac.uk/home/neil.crickmore/Bt/index.html.

Entwistle, P., Bailey, M. J., Cory, J. & Higgs, S. (eds) 1993 *Bacillus thuringiensis: an environmental biopesticide.* New York: Wiley.

Ferré, J., Real, M. D., Van Rie, J., Jansens, S. & Peferoen, M. 1991 Resistance to the *Bacillus thuringiensis* bioinsecticide in a field population of *Plutella xylostella* is due to a change in a midgut membrane receptor. *Proc. Natn. Acad. Sci. USA* **88**, 5119–5123.

Ferré, J., Escriche, B., Bel, Y. & Van Rie, J. 1995 Biochemistry and genetics of insect resistance to *Bacillus thuringiensis* insecticidal crystal proteins. *FEMS Microbiol. Lett.* **132**, 1–7.

ffrench-Constant, R. H., Steichen, J., Rocheleau, T. A., Aronstein, K. & Roush, R. T. 1993 A single amino acid substitution in a gamma-aminobutyric acid subtype A receptor locus associated with cyclodiene insecticide resistance in *Drosophila* populations. *Proc. Natn. Acad. Sci. USA* **90**, 1957–1961.

Gill, S. S., Cowles, E. A. & Pietrantonio, P. V. 1992 The mode of action of *Bacillus thuringiensis* endotoxins. *A. Rev. Entomol.* **37**, 615–636.

Gould, F. 1998 Sustainability of transgenic insecticidal cultivars: integrating pest genetics and ecology. *A. Rev. Entomol.* **43**, 701–726.

Gould, F., Martinez-Ramirez, A., Anderson, A., Ferré, J., Silva, F. J. & Moar, W. J. 1992 Broad-spectrum resistance to *Bacillus thuringiensis* toxins in *Heliothis virescens. Proc. Natn. Acad. Sci. USA* **89**, 7986–7990.

Gould, F., Anderson, A., Reynolds, A., Bumgarner, L. & Moar, W. 1995 Selection and genetic analysis of a *Heliothis virescens* (Lepidoptera: Noctuidae) strain with high levels of resistance to *Bacillus thuringiensis* toxins. *J. Econ. Entomol.* **88**, 1545–1559.

Gould, F., Anderson, A., Jones, A., Sumerford, D., Heckel, D. J., Lopez, J., Micinski, S., Leonard, R. & Laster, M. 1997 Initial frequency of alleles for resistance to *Bacillus thuringiensis* toxins in field populations of *Heliothis virescens. Proc. Natn. Acad. Sci. USA* **94**, 3519–3523.

Granero, F., Ballester, V. & Ferré, J. 1996 *Bacillus thuringiensis* crystal proteins CrylAb and CrylF share a high affinity binding site in *Plutella xylostella* (L.). *Biochem. Biophys. Res. Commun.* **224**, 779–783.

Grochulski, P., Masson, L., Borisova, S., Pusztai-Carey, M., Schwartz, J. L., Brousseau, R. & Cygler, M. 1995 *Bacillus thuringiensis* CryIA(a) insecticidal toxin: crystal structure and channel formation. *J. Molec. Biol.* **254**, 447–464.

Heckel, D. G., Gahan, L. C., Gould, F. & Anderson, A. 1997 Identification of a linkage group with a major effect on resistance to *Bacillus thuringiensis* CryIAc endotoxin in the tobacco budworm (Lepidoptera: Noctuidae). *J. Econ. Entomol.* **90**, 75–86.

Lee, M. K., Rajamohan, F., Gould, F. & Dean, D. H. 1995 Resistance to *Bacillus thuringiensis* CryIA δ-endotoxin in a laboratory-selected *Heliothis virescens* strain is related to receptor alteration. *Appl. Environ. Microbiol.* **61**, 3836–3842.

Li, J., Carroll, J. & Ellar, D. J. 1991 Crystal structure of insecticidal-endotoxin from *Bacillus thuringiensis* at 2.5Å resolution. *Nature* **353**, 815–821.

Liu, Y.-B. & Tabashnik, B. E. 1997 Inheritance of resistance to *Bacillus thuringiensis* toxin CryIC in diamondback moth. *Appl. Environ. Microbiol.* **63**, 2218–2223.

McGaughey, W. H. 1985 Insect resistance to the biological insecticide *Bacillus thuringiensis*. *Science* **229**, 193–195.

McGaughey, W. H. & Johnson, D. E. 1992 Indianmeal moth (Lepidoptera: Pyralidae) resistance to different strains and mixtures of *Bacillus thuringiensis*. *J. Econ. Entomol.* **85**, 1594–1600.

McGaughey, W. H. & Johnson, D. E. 1994 Influence of crystal protein composition of *Bacillus thuringiensis* strains on cross-resistance in Indianmeal moth (Lepidoptera: Pyralidae). *J. Econ. Entomol.* **87**, 535–540.

McGaughey, W. H. & Whalon, M. E. 1992 Managing insect resistance to *Bacillus thuringiensis* toxins. *Science* **258**, 1451–1455.

MacIntosh, S., Stone, T., Jokerst, R. & Fuchs, R. 1991. Binding of *Bacillus thuringiensis* proteins to a laboratory-selected line of *Heliothis virescens*. *Proc. Natn. Acad. Sci. USA* **88**, 8930–8933.

McKenzie, J. A. & Batterham, P. 1994 The genetic, molecular and phenotypic consequences of selection for insecticide resistance. *Trends Ecol. Evol.* **9**, 166–169.

Mellon, M. & Rissler, J. (eds) 1998 *Now or never: serious plans to save a natural pest control.* Cambridge, MA: Union of Concerned Scientists Publications.

Moar, W. J., Pusztai-Carey, M., van Faassen, H., Bosch, D., Frutos, R., Rang, C., Luo, K. & Adang, M. J. 1995 Development of *Bacillus thuringiensis* CryIC resistance by *Spodoptera exigua* (Hübner) (Lepidoptera: Noctuidae). *Appl. Environ. Microbiol.* **61**, 2086–2092.

Oppert, B., Kramer, K. J., Beeman, R. W., Johnson, D. & McGaughey, W. H. 1997 Proteinase-mediated insect resistance to *Bacillus thuringiensis* toxins. *J. Biol. Chem.* **272**, 23473–23476.

Perez, C. J. & Shelton, A. M. 1997 Resistance of *Plutella xylostella* (Lepidoptera: Plutellidae) to *Bacillus thuringiensis* Berliner in Central America. *J. Econ. Entomol.* **90**, 87–93.

Raymond, M., Callaghan, A., Fort, P. & Pasteur, N. 1991 Worldwide migration of amplified insecticide resistance genes in mosquitoes. *Nature* **350**, 151–153.

Sims, S. R. & Stone, T. B. 1991 Genetic basis of tobacco budworm resistance to an engineered *Pseudomonas fluorescens* expressing the δ-endotoxin of *Bacillus thuringiensis* subsp. *kurstaki*. *J. Invert. Pathol.* **57**, 206–210.

Tabashnik, B. E. 1994 Evolution of resistance to *Bacillus thuringiensis*. *A. Rev. Entomol.* **39**, 47–79.

Tabashnik, B. E. 1997 Seeking the root of insect resistance to transgenic plants. *Proc. Natn. Acad. Sci. USA* **94**, 3488–3490.

Tabashnik, B. E., Cushing, N. L., Finson, N. & Johnson, M. W. 1990 Field development of resistance to *Bacillus thuringiensis* in diamondback moth (Lepidoptera: Plutellidae). *J. Econ. Entomol.* **83**, 1671–1676.

Tabashnik, B. E., Finson, N., Groeters, F. R., Moar, W. J., Johnson, M. W., Luo, K. & Adang, M. J. 1994 Reversal of resistance to *Bacillus thuringiensis* in *Plutella xylostella*. *Proc. Natn. Acad. Sci. USA* **91**, 4120–4124.

Tabashnik, B. E., Finson, N., Johnson, M. W. & Heckel, D. G. 1995 Prolonged selection affects stability of resistance to *Bacillus thuringiensis* in diamondback moth (Lepidoptera: Plutellidae). *J. Econ. Entomol.* **88**, 219–224.

Tabashnik, B. E., Malvar, T., Liu, Y.-B., Finson, N., Borthakur, D., Shin, B.-S., Park, S.-H., Masson, L., de Maagd, R. A. & Bosch, D. 1996 Cross-resistance of the diamondback moth indicates altered interactions with domain II of *Bacillus thuringiensis* toxins. *Appl. Environ. Microbiol.* **62**, 2839–2844.

Tabashnik, B. E., Liu, Y.-B., Finson, N., Masson, L. & Heckel, D. G. 1997a One gene in diamondback moth confers resistance to four *Bacillus thuringiensis* toxins. *Proc. Natn. Acad. Sci. USA* **94**, 1640–1644.

Tabashnik, B. E., Liu, Y.-B., Malvar, T., Heckel, D. G., Masson, L., Ballester, V., Granero, F., Mensua, J. L. & Ferré, J. 1997b Global variation in the genetic and biochemical basis of diamondback moth resistance to *Bacillus thuringiensis*. *Proc. Natn. Acad. Sci. USA* **94**, 12780–12785.

Talekar, N. S. & Shelton, A. M. 1993 Biology, ecology, and management of diamondback moth. *A. Rev. Entomol.* **38**, 275–301.

Tang, J. D., Shelton, A. M., Van Rie, J., de Roeck, S., Moar, W. J., Roush, R. T. & Peferoen, M. 1996 Toxicity of *Bacillus thuringiensis* spore and crystal protein to resistant diamondback moth (*Plutella xylostella*). *Appl. Environ. Microbiol.* **62**, 564–569.

Tang, J. D., Gilboa, S., Roush, R. T. & Shelton, A. M. 1997 Inheritance, stability, and lack-of-fitness costs of field-selected resistance to *Bacillus thuringiensis* in diamondback moth (Lepidoptera: Plutellidae) from Florida. *J. Econ. Entomol.* **90**, 732–741.

Van Rie, J., McGaughey, W. H., Johnson, D. E., Barnett, M. D. & Van Mellaert, H. 1990 Mechanism of insect resistance to the microbial insecticide *Bacillus thuringiensis*. *Science* **247**, 72–74.

Wadman, M. 1997 Dispute over insect resistance to crops. *Nature* **388**, 817.

Wright, D. J., Iqbal, M., Granero, F. & Ferré, J. 1997 A change in a single midgut receptor in the diamondback moth (*Plutella xylostella*) is only in part responsible for field resistance to *Bacillus thuringiensis* subsp. *kurstaki* and *B. thuringiensis* subsp. *aizawai*. *Appl. Environ. Microbiol.* **63**, 1814–1819.

Zhao, J.-Z., Zhu, G.-R., Zhu, Z.-L. & Wang, W.-Z. 1993 Resistance of diamondback moth to *Bacillus thuringiensis* in China. *Resist. Pest Mgmt* **5**, 11–12.

Challenges with managing insecticide resistance in agricultural pests, exemplified by the whitefly *Bemisia tabaci*

I. Denholm[1]*, M. Cahill[1], T. J. Dennehy[2] and A. R. Horowitz[3]

[1]*Department of Biological and Ecological Chemistry, IACR-Rothamsted, Harpenden, Herts AL5 2JQ, UK*
[2]*Department of Entomology, University of Arizona, Tucson, AZ 85721, USA*
[3]*Department of Entomology, Institute of Plant Protection, Volcani Center, PO Box 6, Bet Dagan 50250, Israel*

For many key agricultural pests, successful management of insecticide resistance depends not only on modifying the way that insecticides are deployed, but also on reducing the total number of treatments applied. Both approaches benefit from a knowledge of the biological characteristics of pests that promote or may retard the development of resistance. For the whitefly *Bemisia tabaci* (Gennadius), these factors include a haplodiploid breeding system that encourages the rapid selection and fixation of resistance genes, its breeding cycle on a succession of treated or untreated hosts, and its occurrence on and dispersal from high-value crops in greenhouses and glasshouses. These factors, in conjunction with often intensive insecticide use, have led to severe and widespread resistance that now affects several novel as well as conventional control agents. Resistance-management strategies implemented on cotton in Israel, and subsequently in southwestern USA, have nonetheless so far succeeded in arresting the resistance treadmill in *B. tabaci* through a combination of increased chemical diversity, voluntary or mandatory restrictions on the use of key insecticides, and careful integration of chemical control with other pest-management options. In both countries, the most significant achievement has been a dramatic reduction in the number of insecticide treatments applied against whiteflies on cotton, increasing the prospect of sustained use of existing and future insecticides.

Keywords: *Bemisia tabaci*; cotton; ecological genetics; haplodiploidy; resistance management; whitefly

1. INTRODUCTION

Managing resistance to insecticides entails finding practical solutions to problems posed by the adaptability of insect pests. The most obvious and politically seductive option is to abandon insecticides and switch entirely to cultural and/or biological control methods. Unfortunately, this is rarely a feasible proposition. The first and unquestionably most productive line of attack is none the less to minimize use of chemicals to the point at which various operational tactics for reducing or diversifying selection pressure (reviewed by Roush & McKenzie 1987; Roush 1989; Tabashnik 1990; Denholm & Rowland 1992) can be brought into play.

The maximum number of insecticide applications that can be accommodated whilst suppressing resistance will depend on several interacting factors including the genetic and ecological attributes of specific pests, the number of non-cross-resisted chemicals available, and the type of management tactics adopted. As a general rule, pests whose biology renders them most prone to developing resistance are those for which restrictions on insecticide exposure should be most stringent. The basic dilemma is that these are often the species whose biology also renders them most abundant and damaging, and therefore

primary targets of insecticide treatments. For this reason, they may also be ones in which resistance is already well advanced, diminishing the supply of effective compounds and placing new insecticides under severe threat of overuse and hence resistance from the outset (Cahill & Denholm 1998; Denholm *et al.* 1998).

The challenges of combating resistance within such constraints are well exemplified by the cotton, tobacco or sweet-potato whitefly, *Bemisia tabaci* (Gennadius). *B. tabaci* is now one of the most serious agricultural pests worldwide. It is highly polyphagous, attacking numerous field crops throughout the tropics and sub-tropics, and has recently become a significant pest of protected horticulture in temperate regions (Denholm *et al.* 1996). It reduces crop yields by direct feeding damage, through honeydew production that contaminates produce, and by transmitting more than 60 plant viruses (Bedford *et al.* 1994). In many cropping systems, the capacity of this species to evolve resistance has precipitated a classic treadmill of increasing numbers of applications and rapid depletion of effective control agents (Dittrich *et al.* 1990; Byrne *et al.* 1992; Horowitz *et al.* 1994; Denholm *et al.* 1996; Dennehy & Williams 1997). In this paper, we explore some of the biological characteristics of *B. tabaci* that promote resistance in agricultural and horticultural systems, and consider their implications for resistance to conventional and new insecticides. More optimistically,

*Author for correspondence (ian.denholm@bbsrc.ac.uk).

we then describe two recent control strategies involving insecticides that have not only mitigated resistance problems in the short-term, but have also succeeded in reducing chemical use to the point at which long-term, sustainable resistance management becomes a viable proposition.

The taxonomic treatment of pest populations of *Bemisia* is currently controversial and inconsistent. In this paper, we continue to regard all populations exhibiting published morphological traits of *B. tabaci* as belonging to this species, including ones recently described as a distinct species, *B. argentifolii* (Perring *et al.* 1993; Bellows *et al.* 1994).

2. FACTORS INFLUENCING SELECTION RATES

(a) *Influence of breeding system*

Like all members of the Aleyrodidae, *B. tabaci* has long been assumed to possess a breeding system based on haplodiploidy, whereby males are produced uniparentally from unfertilized, haploid eggs, and females are produced biparentally from fertilized, diploid eggs (White 1973). The ability of unmated females to generate viable male progeny (Byrne & Devonshire 1996) identifies this system as one of true arrhenotoky, distinct from 'parahaploidy' (*sensu* Hoy 1979), 'psuedo-arrhenotoky' (*sensu* Schulten 1985) and 'functional haplodiploidy' (*sensu* Brun *et al.* 1995). In the latter three cases, mating is essential for offspring production but one set of male chromosomes is eliminated or inactivated during early development. Until recently, however, there was no direct genetic evidence for haplodiploidy in *B. tabaci*. Through crossing experiments between whitefly strains differing in polymorphic esterase and acetylcholinesterase alleles, Byrne & Devonshire (1996) demonstrated unequivocally the failure of male progeny to express paternal markers, rendering them hemizygous for one or other maternal allele. Blackman & Cahill (1997) have since published karyotypes for several strains of *B. tabaci*, showing males to possess only half ($n=10$) the female complement ($2n=20$) of chromosomes.

From the standpoint of selecting for resistance, the primary consequence of haplodiploidy is that resistance genes arising by mutation are exposed to selection from the outset in hemizygous males, irrespective of intrinsic dominance or recessiveness. The ability of this to accelerate resistance development has been noted by several authors (Helle 1968; Havron *et al.* 1987; Brun *et al.* 1995; Caprio & Hoy 1995), and is reinforced by results of a simple simulation model (figure 1). This model assumes selection to occur before random mating between genotypes, a 1:1 sex-ratio among progeny, and the fitness of hemizygous male genotypes under insecticide exposure to be identical to that of homozygous female counterparts. The latter may not apply to all resistance mechanisms, e.g. mechanisms involving enhanced detoxification in which dosage effects may occur, but this has relatively little influence on overall conclusions. Whether a resistance allele is effectively dominant (figure 1*a*), semi-dominant (figure 1*b*), or recessive (figure 1*c*), resistance develops at a similar rate under haplodiploidy, whereas recessiveness can cause substantial delays in corresponding diploid populations. When resistance alleles are

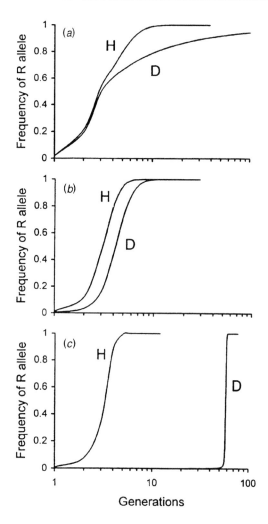

Figure 1. Simulated selection for resistance in haplodiploid (H) and diploid (D) species when resistance is (*a*) fully dominant, (*b*) semi-dominant, or (*c*) fully recessive. The initial resistance (R) allele frequency is 0.001. Fitness values assigned to SS and RR genotypes are 0.05 and 0.9, respectively. Fitness of the RS heterozygote is 0.9 (dominant), 0.3 (semi-dominant) or 0.05 (recessive). Fitness values for hemizygous S and R genotypes are assumed to be identical to those of their SS and RR counterparts.

dominant or nearly so (see, for example, figure 1*a*), another consequence of hemizygosity is to speed up their rate of fixation, because alleles for susceptibility are no longer shielded in males.

The potential of haplodiploidy to promote rapid selection of resistance to high frequencies is still under appreciated, despite the prevalence of this breeding system in many key arthropod pests, including whiteflies, spider mites and phytophagous thrips (White 1973; Schulten 1985; Havron *et al.* 1987). It undoubtedly contributes to the capacity of *B. tabaci* to evolve resistance rapidly to newer insecticides, especially in enclosed environments (see below), and has profound implications for the sustainability of any transgenes incorporated into crop plants to control such species. Maintenance of high toxin expression levels to exploit the presumed lower resistance of heterozygotes compared with homozygotes is currently the most favoured option for delaying resistance to transgenic crops deployed against lepidopteran and coleopteran species (see, for example, Roush 1997).

(b) *Ecology in field crops*

For highly polyphagous species such as *B. tabaci*, interactions between pest ecology and resistance can be subtle and complex, reflecting the seasonality and relative abundance of treated and untreated hosts, and patterns of migration between hosts at different times of the year. There are still very few agricultural systems for which these interactions have been resolved sufficiently to understand their influence on resistance development. The best understood systems of all are probably those involving cotton, melons and vegetables in the southwestern deserts of the United States, where long-term research into whitefly ecology and migration is complemented by extensive monitoring of changes in resistance levels.

The key feature of these systems is the continuous availability of suitable host plants, enabling *B. tabaci* populations to develop actively throughout the year (Coudriet *et al.* 1986; Byrne *et al.* 1990). Typically, whiteflies overwinter on weeds, lettuce and brassica crops, migrating short distances to colonize melons and cotton in the spring and summer, respectively (Byrne & Blackmer 1996). Both of the latter crops are treated frequently with insecticides, unlike alfalfa, which is present all year and seldom sprayed, therefore providing a refuge for insects carrying susceptibility genes.

Regional variation in these production systems has profound implications for the speed at which resistance is selected and for the effectiveness of resistance-management recommendations. In the Imperial Valley of southeastern California, spring melons remain an important crop but the acreage of cotton has declined dramatically, now accounting for only 6000–8000 acres (1 acre = 0.404 686 ha) compared with *ca.* 200 000 acres of unsprayed alfalfa (Castle *et al.* 1996*a,b*). Regular monitoring of whitefly resistance since 1993 by means of insecticide-impregnated sticky cards has shown that, although resistance has generally increased each season as a consequence of insecticide use on spring melons, it they have tended to decline subsequently on cotton, owing partly to extensive immigration of susceptible whiteflies into cotton from adjacent, untreated hosts. As a result, there has been no overall increase in the severity of resistance problems over successive seasons (Castle *et al.* 1996*a*; S. J. Castle, personal communication). Interactions between cropping patterns, chemical use and the bionomics of *B. tabaci* have in this case led *de facto* to preserving susceptibility to several widely used insecticides.

In the neighbouring state of Arizona, resistance monitoring has disclosed substantial geographical variation in resistance levels that can also be reconciled with the bionomics of *B. tabaci*. Before the introduction of a resistance-management strategy for cotton in 1996 (see below), resistance to insecticides (e.g. pyrethroids) used most frequently on that crop was consistently more severe in central Arizona, where cotton constitutes approximately two-thirds of the acreage of cultivated whitefly hosts, than in southwestern Arizona, where it accounts for less than one-quarter of the total acreage (Dennehy *et al.* 1996; Williams *et al.* 1997). Although unsprayed alfalfa

accounts for approximately one-quarter of the available whitefly hosts in both areas, differences in the proportion of cotton compared with alfalfa are hypothesized to have accounted for this regional variation in pyrethroid resistance. As a result, the risk of resistance to newer insecticides (e.g. insect growth regulators) used on cotton appears greater in the centre than in the southwest of the state (Denholm *et al.* 1998). Conversely, risks of resistance to compounds (e.g. imidacloprid) used primarily on vegetables and melons appear greater in the southwest, where these crops are proportionally much more abundant than elsewhere in Arizona (Williams *et al.* 1997).

(c) *Ecology in protected crops*

Enclosed environments such as greenhouses and glasshouses, which restrict immigration and escape from insecticide exposure under climatic regimes favouring rapid and uninterrupted population growth, are widely recognized as providing near-ideal conditions for selecting resistance genes (see, for example, Parrella 1987; Sanderson & Roush 1995; Denholm *et al.* 1998). In the case of *B. tabaci* and other species that transmit plant viruses or cause cosmetic damage to high-value ornamental or edible produce, these problems are accentuated by very low or even zero pest tolerance thresholds that promote frequent spraying and hence intensify selection pressures. As a consequence, these environments have historically proved potent sources of novel resistance genes, and it is no coincidence that resistance of *B. tabaci* to several newer insecticides was first reported in protected horticulture (Horowitz *et al.* 1994; Cahill *et al.* 1996*a,b*).

Selection of resistance in greenhouses and glasshouses not only generates control problems *in situ*, but can have wider implications due to gene flow from these sites. On a local scale, active migration of insects from protected environments to adjacent field crops has the potential to infuse outdoor populations with resistance genes and thereby accelerate their selection to damaging frequencies. To our knowledge this has not been demonstrated directly for *B. tabaci*, but it constitutes a tangible threat where outdoor and protected hosts are grown in close proximity. In contrast, the risk of large-scale, inadvertent movement of resistant insects between countries or even continents via the international trade in ornamental plants is clearly apparent from the high levels of resistance found in individuals of *B. tabaci* newly imported into northern European glasshouses (Cahill *et al.* 1994; Denholm *et al.* 1996). In this respect, *B. tabaci* provides an even more clear-cut example of resistance genes being spread by human agency than that inferred from molecular studies on mosquitoes (Raymond *et al.* 1991).

3. IMPLICATIONS FOR RESISTANCE

(a) *Conventional insecticides*

In many parts of the world, the above factors combined with often intensive insecticide use have led to strong resistance encompassing the great majority of chemical classes (Dittrich *et al.* 1990; Byrne *et al.* 1992; Cahill *et al.* 1995, 1996*c*; Denholm *et al.* 1996; Dennehy & Williams 1997). Resistance in *B. tabaci* is known to be multi-factorial,

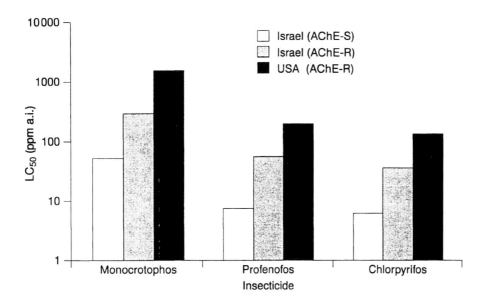

Figure 2. Response of adults of three strains of *Bemisia tabaci* to three organophosphorous insecticides, expressed as LC_{50} values calculated by probit analysis from results of leaf-dip bioassays.

based on both enhanced detoxification of insecticides and modifications to three of their major target proteins: acteylcholinesterase (AChE), targeted by organophosphates (OPs) and carbamates (Byrne & Devonshire 1993; Anthony *et al.* 1998), the GABA-gated chloride-ion channel, targeted by cyclodienes (Anthony *et al.* 1995), and the voltage-sensitive sodium channel involved in knockdown resistance (kdr) to pyrethroids (M. S. Williamson, personal communication). Modified AChE involves at least two allelic variants with different insensitivity profiles (Byrne & Devonshire 1997), and it is possible that kdr resistance in *B. tabaci* will also prove multi-allelic, as in other important pests (Martinez-Torres *et al.* 1997).

In comparison with species such as houseflies (see, for example, Sawicki 1973) and peach–potato aphids (Field *et al.* 1997), there has been relatively little progress towards isolating individual mechanisms in *B. tabaci* to determine their contribution, singly or in combination, to overall resistance phenotypes. Work on the genetics of resistance is complicated by many field populations appearing homozygous for resistance genes (Denholm *et al.* 1996), and by reproductive barriers between *B. tabaci* 'biotypes' (see, for example, Bedford *et al.* 1994). By rearing single-pair progeny from an Israeli population polymorphic for insecticide-sensitive (AChE-S) and insensitive (AChE-R) alleles, Byrne *et al.* (1994) were able to isolate strains homozygous for each of these alleles in a common genetic background. Differences in LC_{50} values of these strains for three OP insecticides (six- to ninefold; figure 2) could therefore be attributed to this target-site mechanism alone. However, several other strains homozygous for the same resistance allele, including that from the USA included in figure 2, have shown significantly higher resistance than the Israeli one. Such strains must possess at least one additional mechanism, most likely based on enhanced detoxification of OPs prior to their reaching the insensitive target site.

The implications of multiple mechanisms for resistance management are exemplified well by events on cotton in the southwestern USA. During the early 1990s, one response to increasing resistance in *B. tabaci* was to screen numerous combinations of products for possible synergistic effects (see, for example, Akey *et al.* 1993; Wolfenbarger & Riley 1994). Some mixtures of pyrethroids and OPs, especially of fenpropathrin and acephate, proved remarkably effective in this respect as shown by laboratory bioassay data for a whitefly strain collected from Safford, Arizona, in 1994 (figure 3a). Despite immunity to acephate and very high resistance to fenpropathrin applied alone, co-application of a fixed concentration of 1000 ppm acephate (non-toxic in its own right) with varying concentrations of fenpropathrin increased the toxicity of the pyrethroid by over 1000-fold, to a level at which field efficacy was greater than ever observed in Arizona for fenpropathrin alone. Although the biochemical basis of this extreme synergy has not been fully resolved, it most probably reflects inhibition by acephate of metabolic enzymes conferring resistance to fenpropathrin and other pyrethroids applied singly.

Based on this finding, tank mixes of pyrethroids and OPs became widely adopted for controlling *B. tabaci*, and were by far the most commonly used treatments on Arizona cotton during 1994 and 1995. By the end of the 1995 season, however, such mixtures had failed to control *B. tabaci* throughout much of central Arizona (Dennehy & Williams 1997). Laboratory tests against adults collected during 1995 from the Maricopa Agricultural Centre (Simmons & Dennehy 1996; summarized in figure 3b) confirmed that repeated use of synergized pyrethroids between July and September had led to a gradual loss of synergism of fenpropathrin by acephate. This resistance declined sharply between the 1995 and 1996 seasons, but a single treatment with the mixture between July and August 1996 caused it to increase again.

Resistance to synergized pyrethroids could reflect the selection of a new, non-synergizable resistance mechanism (e.g. target-site kdr resistance) or a modification of one already present; this is currently being investigated. Its appearance demonstrates that even tactics optimized to contend with existing resistance mechanisms can be rapidly compromised by the genetic plasticity of insect pests. Owing to the apparent instability of this resistance,

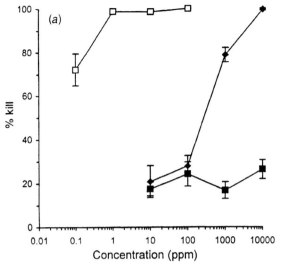

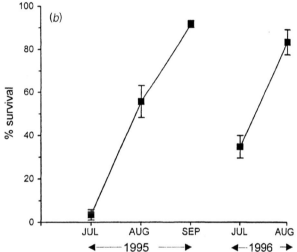

Figure 3. (*a*) Response of *B. tabaci* collected from cotton in Safford, Arizona, in 1994 to the pyrethroid fenpropathrin, the OP acephate, or a combination of both chemicals. Filled squares (■), varying concentrations of acephate alone; diamonds (◆), varying concentrations of fenpropathrin alone; open squares (□), varying concentrations of fenpropathrin plus a fixed concentration of 1000 ppm acephate. (*b*) Changes in response of *B. tabaci* on cotton in Maricopa, Arizona, during 1995 and 1996 to a combination of 10 ppm fenpropathrin and 1000 ppm acephate. All results relate to leaf-dip bioassays against adult whiteflies.

pyrethroid/OP combinations remain a viable management option if used with extreme moderation. However, their importance in this region has diminished since the introduction of more novel control agents that now underpin the management of resistance in *B. tabaci*.

(b) *Newer insecticides*

With the majority of older insecticides affected by resistance to varying degrees, compounds with novel or less exploited modes of action are assuming major importance as components of whitefly control programmes (Horowitz & Ishaaya 1994; Denholm *et al.* 1996; Cahill & Denholm 1998; Ishaaya & Horowitz

1998). Most of these offer the considerable advantage of being unaffected by existing resistance mechanisms, and many also exhibit more favourable environmental profiles than broad-spectrum agents employed in the past. However, all come with intrinsic resistance risks, and there is already ample evidence of how these may be accentuated by the genetics and life history of *B. tabaci*.

The chemicals probably attracting most attention in this respect are the chloronicotinyls or neonicotinoids, whose forerunner, imidacloprid, is now in widespread use against whiteflies and other pests including aphids and the Colorado beetle, *Leptinotarsa decemlineata* (Say) (Elbert *et al.* 1996). Other members of this class are currently being released or in late stages of development (Casida 1998). Their efficacy both as highly persistent systemic treatments and as shorter-lived foliar sprays offers outstanding versatility, but also renders them particularly prone to overuse. Their structural similarity, coupled with a likely common target site, nicotinic acetylcholine receptors in the post-synaptic region of insect nerves (Bai *et al.* 1991), also raises the spectre of cross-resistance affecting the group as a whole (Elbert *et al.* 1996; Cahill & Denholm 1998). The risk of resistance is therefore considerable, and reinforced by the speed with which resistance to imidacloprid has developed in *B. tabaci* under sustained exposure to this chemical in the laboratory (Prabhakar *et al.* 1997) and in greenhouses in southern Europe (Cahill *et al.* 1996*b*). Given the current scale of imidacloprid use, further resistance outbreaks seem inevitable and it is essential to exploit these for defining conditions under which chloronicotinyls might be used sustainably. Large-scale laboratory experiments with Spanish strains of *B. tabaci* have demonstrated substantial differences in the phenotypic expression of imidacloprid resistance under systemic and foliar treatment regimes, but also highlighted the dangers of unrestricted, successive exposure to both types of application, especially in enclosed environments (Cahill & Denholm 1998, unpublished data).

Despite their key role in sustaining some current resistance management strategies for whiteflies on cotton (see below), the insect growth regulators (IGRs) buprofezin and pyriproxyfen are also proving vulnerable to resistance by *B. tabaci*. Although both chemicals act primarily against immature stages of whiteflies, they possess distinct modes of action and are therefore very unlikely to be affected by the same resistance mechanism. Buprofezin inhibits chitin synthesis and results in nymphal death during ecdysis (Uchida *et al.* 1985), whereas pyriproxyfen is a juvenile hormone mimic interrupting nymphal and pupal development, but also suppressing egg hatch by direct exposure of eggs or transovarially via the treatment of adult females (Ishaaya & Horowitz 1992). These differences, coupled with a high degree of species selectivity, make them ideally suited as rotation partners in control programmes that place emphasis on the preservation of natural enemies of *B. tabaci* and co-existing pest species (Horowitz *et al.* 1994).

As with imidacloprid, resistance to both IGRs first became apparent under the intensive selection operating in protected environments. Resistance to buprofezin was first detected in glasshouses in The Netherlands, but has since been demonstrated elsewhere in northern Europe

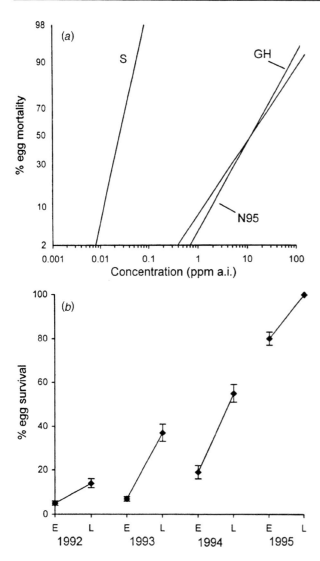

Figure 4. (*a*) Probit lines summarizing the response of three strains of *B. tabaci* from Israel to the IGR pyriproxyfen. S refers to the reference susceptible strain; GH refers to the insects collected from roses in a greenhouse in January 1992; N95 refers to the insects collected from cotton at Nachshon in central Israel in late 1995. (*b*) Changes in the response of *B. tabaci* from early-season (E) and late-season (L) season cotton at Nachshon to a discriminating concentration of 0.16 ppm pyriproxyfen between 1992 and 1995. All results relate to a leaf-dip bioassay exposing adults to the toxin and measuring subsequent egg hatch.

and in greenhouses in Spain and Israel (Horowitz & Ishaaya 1994; Cahill *et al.* 1996*a*). In many cases, the appearance of resistance could be attributed to repeated applications of this chemical with no regard to resistance management, although in the UK it preceded the official approval of buprofezin and unquestionably resulted from the accidental importation of resistant insects on plant produce (Cahill *et al.* 1996*a*). Buprofezin is still less widely used in the open field and resistance has been slower to develop. However, monitoring of the susceptibility of *B. tabaci* on cotton in Israel, where this chemical has been restricted to a single application per year since its introduction in 1989, has shown small increases in LC$_{50}$

values consistent with the presence of resistant individuals at some localities (Horowitz *et al.* 1994).

The history of pyriproxyfen use against *B. tabaci* in Israel provides possibly the most striking example of genetic, ecological and operational factors conspiring to promote resistance, despite conscious attempts to prevent this occurring. Its introduction in Israel in 1991 represented the first ever commercial use against white-flies of a compound developed primarily to control household and public health pests. Within one year of its introduction, high resistance was demonstrated in some ornamental greenhouses after successive applications (Horowitz & Ishaaya 1994; Horowitz *et al.* 1994). The maximum resistance recorded was *ca.* 550-fold at LC$_{50}$, from a rose greenhouse that had previously been sprayed only three times with this chemical (figure 4*a*). The linearity of dose–response data obtained from bioassays at this stage implied a high degree of homozygosity for one or more pyriproxyfen-resistance genes.

As with buprofezin, pyriproxyfen was released for use on Israeli cotton with a restriction to one application per season, in order to prevent or delay the development of resistance. Despite excellent compliance with this recommendation, resistance was first detected on cotton as early as 1992 in some localities, where it has since undergone a gradual increase in frequency within and between seasons. The most dramatic changes were recorded in the vicinity of Nachshon in the Ayalon Valley in central Israel, where egg survival at a diagnostic concentration of 0.16 ppm showed a stepwise increase between 1992 and 1995 (figure 4*b*). By late 1995, the fitted dose–response probit line for Nachshon insects was identical to that for the most resistant greenhouse population (figure 4*a*). The reasons why a maximum of only five treatments spaced one year apart should have selected so effectively for resistance are still unclear. Because Nachshon is remote from areas of intensive protected horticulture, contamination by resistant insects from greenhouses seems unlikely. The more plausible explanation invokes a combination of adverse factors discussed earlier in this paper: haplo-diplody negating gene dominance and promoting rapid fixation of resistance alleles, and the 'bottlenecking' of whiteflies on irrigated and treated cotton in summer despite an apparent abundance of alternative hosts at other times of the year. A comparison of the ecology of *B. tabaci* at Nachshon with that in other areas where pyriproxyfen resistance has been slower to develop could help considerably with anticipating the risk of resistance to future products and in modifying use recommendations accordingly.

4. MANAGEMENT OF RESISTANCE

Many of the biological characteristics of *B. tabaci* that contribute to its pest status and promote resistance development (e.g. breeding system, migratory ability and polyphagy) cannot be manipulated or controlled directly. In most agricultural systems, however, alteration of agronomic practices, such as the timing and placement of crops, could prove the most effective approach of all in reducing whitefly abundance and hence the need for excessive and continuous reliance on insecticide applications (Byrne & Blackmer 1996; Denholm *et al.*

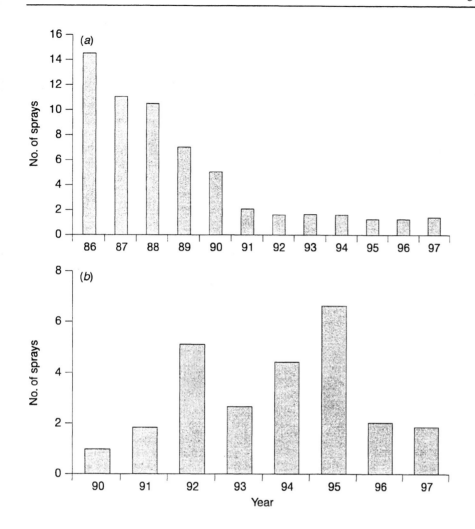

Figure 5. Average number of insecticide sprays applied over successive years against *Bemisia tabaci* on cotton in (*a*) Israel and (*b*) Arizona (data on the latter extracted from Ellsworth (1998)).

1996). In protected horticulture, the use of 'crop-free' periods, although conflicting with many crop-production schedules, is the simplest way to achieve the same objective (Sanderson & Roush 1995). A second, broad approach entails diversifying control tactics to the greatest extent possible, reserving chemicals as the last line of defence. The planting of pest-or virus-tolerant cultivars (see, for example, Nateshan *et al.* 1996), augmentive releases of natural enemies (Hoelmer 1996), application of mineral oils or detergents (Liu & Stansly 1995), and physical barriers to pest movement (Berlinger & Lebuish-Mordechi 1996), all have potential utility in the field and/or protected environments, but require detailed research to tailor them to local conditions. The third set of possible reforms relate to rationalizing and optimizing the way that insecticides are applied, in order to exploit the full diversity of chemicals available and avoid protracted selection for particular resistance mechanisms. The extent to which such reforms can themselves trigger reduced and more sustainable use of insecticides is demonstrated by resistance-management strategies implemented on cotton in Israel and, more recently, in the southwestern USA.

The Israeli strategy was introduced in 1987 in response to increasing failures to control *B. tabaci* with pyrethroids, carbamates and OPs. Its initial purpose was to combat resistance to conventional chemicals through pre-planned rotations of insecticides, but following the release of newer compounds (especially buprofezin in 1989 and pyriproxyfen in 1991), its primary objective has been to preserve susceptibility to these materials by optimizing and restricting their use to a single treatment per year (Horowitz & Ishaaya 1994, 1996; Horowitz *et al.* 1994). A further goal is to conserve whitefly natural enemies of whitefly in cotton by avoiding the use of broad-spectrum agents against other pests in the early stages of the season. Growers are then advised to use more selective chemicals against the American bollworm, *Helicoverpa armigera* Hubner, and to exploit pheromones for controlling the pink bollworm, *Pectinophora gossypiella* (Saunders). The strategy is therefore designed to contend with the entire pest complex and to minimize effects of controlling lepidopteran pests on the buildup and subsequent abundance of *B. tabaci*.

Aided by extensive education programmes and the centralization of pesticide sales, compliance with these recommendations from growers has been outstanding. One of its primary achievements has been a dramatic reduction in the number of insecticide applications against the entire range of cotton pests, but especially against *B. tabaci*. Treatments against the latter now average less than two per year compared with over 14 per year in 1986 (figure 5*a*). This in turn has created an environment favourable for introducing and restricting other novel insecticides (e.g. chloronicotinyls and diafenthiuron) that are now available to cotton growers.

Resistance monitoring is an integral component of the Israeli strategy. In many areas there has been little evidence of reduced susceptibility to IGRs, although the development of resistance to pyriproxyfen in localities such as Nachshon (see above) demonstrates potential ecological 'hot spots' where even a single application per year can select for resistance effectively. In such areas it may prove necessary to impose even more severe temporal or spatial restrictions on particular products in the future. Given the overall reduction in dependence on insecticides for managing *B. tabaci*, the diversity of chemicals now available, and the exceptional degree of cooperation achieved between Israeli researchers, advisors and growers, this is not an impractical proposition.

The situation on cotton in central Arizona in 1995 resembled that in Israel in the late 1980s. Faced with widespread resistance to most conventional insecticides, and increasing control failures with synergistic combinations of pyrethroids and OPs, growers embarked on a resistance treadmill, applying 8–12 whitefly treatments in the most seriously affected areas, and still having cotton discounted by buyers owing to honeydew contamination (Dennehy & Williams 1997).

A speedy response to this crisis was made possible by unprecedented cooperation between local researchers, grower organizations and agrochemical companies, who jointly petitioned the US Environmental Protection Agency (EPA) with a highly unusual request for simultaneous (Section 18) emergency exemption from registration for the two IGRs, buprofezin and pyriproxyfen. The application for a dual exemption was based on resistance-management objectives, and in particular the need to diversify chemical inputs to the greatest extent possible. Supported by data from Israel and the UK demonstrating the resistance risks posed by these compounds, the application also restricted both IGRs to a single application per cotton season.

The resistance-management strategy introduced in Arizona in 1996 divided insecticide use against *B. tabaci* into three stages. The first comprised one or both IGRs, the second non-pyrethroid conventional insecticides, and the third no more than two applications of synergized pyrethroids (Dennehy & Williams 1997; Dennehy & Denholm 1998). As in Israel, however, the effectiveness of IGRs in reducing whitefly densities in cotton have in many cases made pyrethroid treatments unnecessary. Not only have growers adopting the strategy reported greatly improved control of *B. tabaci*, but statewide averages for the number of whitefly treatments per season are estimated to have been reduced from 6.6 in 1995 to 2.0 in 1996 and 1.8 in 1997 (Ellsworth 1998) (figure 5b). In addition, statewide resistance monitoring has shown *B. tabaci* to have regained susceptibility to synergized pyrethroids and key non-pyrethroids in each year since implementation of the strategy (Dennehy & Williams 1997; Dennehy & Denholm 1998).

Despite the success of the Arizona strategy so far, there are no grounds for complacency. In the light of experience with pyriproxyfen in Israel, and the vulnerability of cotton insecticides to resistance under ecological conditions in central Arizona (Dennehy *et al.* 1996), susceptibility of *B. tabaci* to both IGRs is being carefully monitored to anticipate potential resistance problems. When the Section 18 exemption expires at the end of 1998, restrictions on IGRs on cotton will no longer be mandatory, and they are likely to be used on other *B. tabaci* hosts as well (Dennehy & Denholm 1998). There is also concern over the future use of chloronicotinyls on cotton. At present, imidacloprid is applied primarily as a systemic soil treatment to vegetables, and its effectiveness has greatly reduced the number of whiteflies moving from vegetables to cotton early in the year. The anticipated introduction of newer choronicotinyls on cotton would, like the use of IGRs on vegetables, remove the present division of chemistry between crops and place all new insecticides under greater threat from resistance. Avoiding a return to the treadmill of uncontrollable resistance in *B. tabaci* without legislative support is going to place extreme demands on public- and private-sector personnel collectively to limit and harmonize insecticide use in agro-ecosystems in Arizona.

5. CONCLUSIONS

The introduction of resistance-management strategies for *B. tabaci* in Israel and Arizona represents a significant achievement for cotton production in both regions. On a broader scale, these strategies have demonstrated that even spiralling resistance problems in seemingly intractable pest species can be countered effectively by taking full but judicious advantage of the range of chemicals available, and by exploiting the species-selectivity of many new molecules. They have also highlighted the importance of proactive resistance monitoring as a means of evaluating the sustainability of resistance-management recommendations.

In many agricultural systems, the ultimate challenge in combating resistance is not simply to formulate countermeasures but to achieve their implementation amid the commercial and socio-economic pressures that drive the use of pesticides. When reviewing pest management strategies on cotton, Sawicki & Denholm (1987) identified three features likely to assist considerably in this regard. These were: (i) a history of well-documented failures due to resistance; (ii) a good entomological infrastructure for formulating and communicating management recommendations; and (iii) means of regulating the distribution and/or use of agrochemicals. The first two of these unquestionably apply in Israel and Arizona. The third has also been achieved, but in different ways. In Israel, centralization of pesticide sales by the Israel Cotton Board, coupled with discounts on recommended insecticides, has long proved to be a powerful tool for encouraging compliance by growers (Horowitz *et al.* 1994). In Arizona, it is unlikely that the current restrictions on IGRs could have been achieved without the regulatory support provided by the Section 18 exemption, and it remains uncertain whether restrained use of IGRs will be maintained after its expiry (Dennehy & Denholm 1998). Arizona growers recognize the financial and operational benefits that the strategy has already provided, and insecticide manufacturers recognize its importance for the sustainability of their products. Yet it remains to be determined whether Arizona will resort to further regulatory action, such as special label restrictions, or simply promote voluntary compliance with once-per-season use of IGRs once federal approval is granted.

One of the primary threats to managing resistance in *B. tabaci* throughout the world is the continuing, indiscriminate use of insecticides in glasshouses and greenhouses. When reviewing options for resistance management in protected horticulture, Sanderson & Roush (1995) rightly emphasized the extreme importance, and feasibility, of adopting cultural and biological control practices as alternatives to insecticides. This message has been aired repeatedly, but with such limited success to question whether new molecules should be registered for use on protected crops other than within the framework of an established and clearly defined integrated control strategy (Denholm *et al.* 1998). Denying such growers unrestricted access to novel chemistry may seem a radical step, but could have the benefit of forcing the implementation of proven non-chemical control tactics that are often far less exploitable by those attempting to combat resistance development under open-field conditions.

We thank I. Ishaaya for advice, A. R. McCaffery for critically reading the manuscript, and numerous colleagues for valued technical assistance. IACR-Rothamsted receives grant-aided support from the Biotechnology and Biological Sciences Research Council of the UK. Research in Israel is supported by the Israeli Cotton Board and the Chief Scientist of the Ministry of Agriculture. Research in Arizona is supported by the Arizona Cotton Growers Association and Cotton Incorporated.

REFERENCES

Akey, D. H., Henneberry, T. J. & Chu, C. C. 1993 Control studies on field populations of the sweetpotato whitefly, *Bemisia tabaci* in Arizona upland cotton. In *Proceedings of the Beltwide Cotton Production Research Conference*, pp. 675–679. Memphis, TN: National Cotton Council.

Anthony, N. M., Brown, J. K., Markham, P. G. & ffrench-Constant, R. H. 1995 Molecular analysis of cyclodiene resistance-associated mutations among populations of the sweetpotato whitefly *Bemisia tabaci*. *Pestic. Biochem. Physiol.* **51**, 220–228.

Anthony, N. M., Brown, J. K., Feyereisen, R. & ffrench-Constant, R. H. 1998 Diagnosis and characterisation of insecticide-insensitive acetylcholinesterase in three populations of the sweetpotato whitefly *Bemisia tabaci*. *Pestic. Sci.* **52**, 39–46.

Bai, D., Lummis, S. C. R., Leicht, W., Breer, H. & Satelle, D. B. 1991 Actions of imidacloprid and a related nitromethylene on cholinergic receptors of an identified insect motor neurone. *Pestic. Sci.* **33**, 197–204.

Bedford, I. D., Briddon, R. W., Brown, J. K., Rosell, R. C. & Markham, P. G. 1994 Geminivirus transmission and biological characterisation of *Bemisia tabaci* (Gennadius) biotypes from different geographic regions. *Annls Appl. Biol.* **125**, 311–325.

Bellows, T. S., Perring, T. M., Gill, R. J. & Headrick, D. H. 1994 Description of a species of *Bemisia* (Homoptera: Aleyrodidae) infesting North American agriculture. *Annls Entomol. Soc. Am.* **87**, 195–206.

Berlinger, M. & Lebuish-Mardechi, S. 1996 Physical methods for the control of *Bemisia*. In Bemisia *1995: taxonomy, biology, damage, control and management* (ed. D. Gerling & R. T. Mayer), pp. 617–634. Andover, Hants: Intercept.

Blackman, R. L. & Cahill, M. 1997 The karyotype of *Bemisia tabaci* (Hemiptera: Aleyrodidae). *Bull. Entomol. Res.* **87**, 213–215.

Brun, L. O., Stuart, J., Gaudichon, V., Aronstein, K. & ffrench-Constant, R. H. 1995 Functional haplodiploidy: a mechanism for the spread of insecticide resistance in an important international insect pest. *Proc. Natn. Acad. Sci. USA* **92**, 9861–9865.

Byrne, D. N. & Blackmer, J. L. 1996 Examination of short-range migration by *Bemisia*. In Bemisia *1995: taxonomy, biology, damage, control and management* (ed. D. Gerling & R. T. Mayer), pp. 17–28. Andover, Hants: Intercept.

Byrne, F. J. & Devonshire, A. L. 1993 Insensitive acetylcholinesterase and esterase polymorphism in susceptible and resistant populations of the tobacco whitefly *Bemisia tabaci* (Genn.). *Pestic. Biochem. Physiol.* **45**, 34–42.

Byrne, F. J. & Devonshire, A. L. 1996 Biochemical evidence of haplodiploidy in the whitefly *Bemisia tabaci*. *Biochem. Genet.* **34**, 93–107.

Byrne, F. J. & Devonshire, A. L. 1997 Kinetics of insensitive acetylcholinesterases in organophosphate-resistant tobacco whitefly, *Bemisia tabaci* (Gennadius) (Homoptera: Aleyrodidae). *Pestic. Biochem. Physiol.* **58**, 119–124.

Byrne, D. N., Bellows, T. S. & Parrella, M. P. 1990 Whiteflies in agricultural systems. In *Whiteflies: their systematics, pest status and management* (ed. D. Gerling), pp. 227–261. Andover, Hants: Intercept.

Byrne, F. J., Denholm, I., Birnie, L. C., Devonshire, A. L. & Rowland, M. W. 1992 Analysis of insecticide resistance in the whitefly, *Bemisia tabaci*. In *Resistance 91: achievements and developments in combating pesticide resistance* (ed. I. Denholm, A. L. Devonshire & D. W. Hollomon), pp. 165–178. London, UK: Elsevier.

Byrne, F. J., Cahill, M., Denholm, I. & Devonshire, A. L. 1994 A biochemical and toxicological study of the role of insensitive acetylcholinesterase in organophosphorous resistant *Bemisia tabaci* (Homoptera: Aleyrodidae) from Israel. *Bull. Entomol. Res.* **42**, 179–184.

Cahill, M. & Denholm, I. 1998 Managing resistance to the chloronicotinyl insecticides: rhetoric or reality? In *Chloronictinyl insecticides* (ed. J. Casida), ACS Monograph Series no. XX. Washington, DC: American Chemical Society. (In the press.)

Cahill, M., Byrne, F. J., Denholm, I., Devonshire, A. L. & Gorman, K. 1994 Insecticide resistance in *Bemisia tabaci*. *Pestic. Sci.* **42**, 137–139.

Cahill, M., Byrne, F. J., Gorman, K. J., Denholm, I. & Devonshire, A. L. 1995 Pyrethroid and organophosphate resistance in the tobacco whitefly *Bemisia tabaci* (Homoptera: Aleyrodidae). *Bull. Entomol. Res.* **85**, 181–187.

Cahill, M., Jarvis, W., Gorman, K. & Denholm, I. 1996a Resolution of baseline responses and documentation of resistance to buprofezin in *Bemisia tabaci* (Homoptera: Aleyrodidae). *Bull. Entomol. Res.* **86**, 117–122.

Cahill, M., Denholm, I., Gorman, K., Day, S., Elbert, A. & Nauen, R. 1996b Baseline determination and detection of resistance to imidacloprid in *Bemisia tabaci* (Homoptera: Aleyrodidae). *Bull. Entomol. Res.* **86**, 343–349.

Cahill, M., Denholm, I., Byrne, F. J. & Devonshire, A. L. 1996c Insecticide resistance in *Bemisia tabaci*—current status and implications for management. *Proc. 1996 Brighton Crop Protect. Conf.* **1**, 75–80.

Caprio, M. A. & Hoy, M. A. 1995 Premating isolation in a simulation model generates frequency dependent selection and alters establishment rates of resistant natural enemies. *J. Econ. Entomol.* **88**, 205–212.

Casida, J. E. (ed.) 1998 *Chloronictinyl insecticides*. ACS Monograph Series no. XX. Washington, DC: American Chemical Society. (In the press.)

Castle, S. J., Henneberry, T. J., Prabhakar, N. & Toscano, N. C. 1996a Trends in relative susceptibilities of whiteflies to insecticides through the cotton season in the Imperial Valley, California. In *Proceedings of the Beltwide Cotton Production Research Conference*, pp. 1032–1034. Memphis, TN: National Cotton Council.

Castle, S., Henneberry, T., Toscano, N., Prabhakar, N., Birdsall, S. & Weddle, D. 1996*b* Silverleaf whiteflies show no increase in insecticide resistance. *Calif. Agric.* **50**, 18–23.

Coudriet, D. L., Meyerdirk, D. E., Prabhakar, N. & Kishaba, A. N. 1986 Bionomics of sweetpotato whitefly (Homoptera: Aleyrodidae) on weed hosts in the Imperial Valley, California. *Environ. Entomol.* **15**, 1179–1183.

Denholm, I. & Rowland, M. 1992 Tactics for managing pesticide resistance in arthropods: theory and practice. *A. Rev. Entomol.* **37**, 91–112.

Denholm, I., Cahill, M., Byrne, F. J. & Devonshire, A. L. 1996 Progress with documenting and combating insecticide resistance in *Bemisia*. In Bemisia *1995: taxonomy, biology, damage, control and management* (ed. D. Gerling & R. T. Mayer), pp. 577–603. Andover, Hants: Intercept.

Denholm, I., Horowitz, A. R., Cahill, M. & Ishaaya, I. 1998 Management of resistance to novel insecticides. In *Insecticides with novel modes of action: mechanisms and application* (ed. I. Ishaaya & D. Degheele), pp. 260–282. Berlin: Springer.

Dennehy, T. J. & Denholm, I. 1998 Goals, achievements and future challenges of the Arizona whitefly resistance management program. In *Proceedings of the 1998 Beltwide Cotton Production Research Conference*, pp. 68–72. Memphis, TN: National Cotton Council.

Dennehy, T. J. & Williams, L. 1997 Management of resistance in *Bemisia* in Arizona cotton. *Pestic. Sci.* **51**, 398–406.

Dennehy, T. J., Williams, L., Russell, J. S., Li, X. & Wigert, M. 1996 Monitoring and management of whitefly resistance to insecticides in Arizona. In *Proceedings of the 1996 Beltwide Cotton Production Research Conference*, pp. 135–140. Memphis, TN: National Cotton Council.

Dittrich, V., Uk, S. & Ernst, G. H. 1990 Chemical control and insecticide resistance of whiteflies. In *Whiteflies: their systematics, pest status and management* (ed. D. Gerling), pp. 263–285. Andover, Hants: Intercept.

Elbert, A., Nauen, R., Cahill, M., Devonshire, A. L., Scarr, A. W., Sone, S. & Steffens, R. 1996 Resistance management with chloronicotinyl insecticides using imidacloprid as an example. *Pflanzen-Nachrich Bayer* **49**, 5–53.

Ellsworth, P. C. 1998 Whitefly management in Arizona: looking at the whole system. In *Proceedings of the 1998 Beltwide Cotton Production Research Conference*, pp. 65–68. Memphis, TN: National Cotton Council.

Field, L. M., Anderson, A. P., Denholm, I., Foster, S. P., Harling, Z. K., Javed, N., Martinez-Torres, D., Moores, G. D., Williamson, M. S. & Devonshire, A. L. 1997 Use of biochemical and DNA diagnostics for characterising multiple mechanisms of insecticide resistance in the peach–potato aphid, *Myzus persicae* (Sulzer). *Pestic. Sci.* **51**, 283–289.

Havron, A., Rosen, D., Rossler, Y. & Hillel, J. 1987 Selection on the male hemizygous genotype in arrhenotokous insects and mites. *Entomophaga* **32**, 261–268.

Helle, W. 1968 Parthenogenesis and insecticide resistance. *Meded. Rijksfac. Landbouwwet. Gent* **33**, 621–628.

Hoelmer, K. A. 1996 Whitefly parasitoids: can they control field populations of *Bemisia*? In Bemisia *1995 taxonomy, biology, damage, control and management* (ed. D. Gerling & R. T. Mayer), pp. 451–476. Andover, Hants: Intercept.

Horowitz, A. R. & Ishaaya, I. 1994 Managing resistance to insect growth regulators in the sweetpotato whitefly (Homoptera: Aleyrodidae). *J. Econ. Entomol.* **87**, 866–871.

Horowitz, A. R. & Ishaaya, I. 1996 Chemical control of *Bemisia*: management and application. In Bemisia *1995: taxonomy, biology, damage, control and management* (ed. D. Gerling & R. T. Mayer), pp. 537–556. Andover, Hants: Intercept.

Horowitz, A. R., Forer, G. & Ishaaya, I. 1994 Managing resistance in *Bemisia tabaci* in Israel with particular emphasis on cotton. *Pestic. Sci.* **42**, 113–122.

Hoy, M. A. 1979 Parahaploidy of the 'arrhenotokous' predator *Metaseiulus occidentalis* (Acarina: Phytoseiidae) demonstrated by X-irradiation of males. *Entomol. Exp. Appl.* **26**, 97–104.

Ishaaya, I. & Horowitz, A. R. 1992 Novel phenoxy juvenile hormone analog (pyriproxyfen) suppresses embryogenesis and adult emergence of sweetpotato whitefly (Homoptera: Aleyrodidae). *J. Econ. Entomol.* **85**, 2113–2117.

Ishaaya, I. & Horowitz, A. R. 1998 Insecticides with novel modes of action: an overview. In *Insecticides with novel modes of action: mechanisms and application* (ed. I. Ishaaya & D. Degheele), pp. 1–24. Berlin: Springer.

Liu, T. X. & Stansly, P. A. 1995 Toxicity of biorational insecticides to *Bemisia argentifolii* (Homoptera: Aleyrodidae) on tomato leaves. *J. Econ. Entomol.* **88**, 564–568.

Martinez-Torres, D., Devonshire, A. L. & Williamson, M. S. 1997 Molecular studies of knockdown resistance to pyrethoids: cloning of domain II sodium channel gene sequences from insects. *Pestic. Sci.* **51**, 265–270.

Nateshan, H. M., Muniyappa, V., Jalikop, S. H. & Ramappa, H. K. 1996 Resistance of *Lycopersicon* species and hybrids to tomato leaf curl geminiviruses. In Bemisia *1995: taxonomy, biology, damage, control and management* (ed. D. Gerling & R. T. Mayer), pp. 369–377. Andover, Hants: Intercept.

Parrella, M. P. 1987 Biology of *Lyriomyza*. *A. Rev. Entomol.* **32**, 201–224.

Perring, T. M., Cooper, A. D., Rodriguez, R. J., Farrar, C. A. & Bellows, T. S. 1993 Identification of a whitefly species by genomic and behavioural studies. *Science* **259**, 74–77.

Prabhakar, N., Toscano, N. C., Castle, S. J. & Henneberry, T. J. 1997 Selection for resistance to imidacloprid in silverleaf whiteflies from the Imperial Valley and development of a hydroponic bioassay for resistance monitoring. *Pestic. Sci.* **51**, 419–428.

Raymond, M., Callaghan, A., Fort, P. & Pasteur, N. 1991 Worldwide migration of amplified insecticide resistance genes in mosquitoes. *Nature* **350**, 151–153.

Roush, R. T. 1989 Designing resistance management programs: how can you choose? *Pestic. Sci.* **26**, 423–441.

Roush, R. T. 1997 Bt-transgenic crops: just another pretty insecticide or a chance for a new start in resistance management? *Pestic. Sci.* **51**, 328–334.

Roush, R. T. & McKenzie, J. A. 1987 Ecological genetics of insecticide and acaricide resistance. *A. Rev. Entomol.* **32**, 361–380.

Sanderson, J. P. & Roush, R. T. 1995 Management of insecticide resistance in the greenhouse. In *Proc. 11th Conf. Insect Disease Mgmt, Ornamentals* (ed. A. Bishop, M. Hansbeck & R. Lindquist), pp. 18–20. Alexandria, VA: Society of American Florists.

Sawicki, R. M. 1973 Recent advances in the study of the genetics of resistance in the housefly, *Musca domestica*. *Pestic. Sci.* **4**, 501–512.

Sawicki, R. M. & Denholm, I. 1987 Management of resistance to pesticides in cotton pests. *Trop. Pest Mgmt.* **33**, 262–272.

Schulten, G. G. M. 1985 Pseudo-arrhenotoky. In *Spider mites: their biology, natural enemies and control*, vol. 1B (ed. W. Helle & M. W. Sabelis), pp. 67–71. Amsterdam: Elsevier.

Simmons, A. L. & Dennehy, T. J. 1996 Contrasts of three insecticide resistance monitoring methods for whitefly. In *Proceedings of the 1996 Beltwide Cotton Production Research Conference*, pp. 748–752. Memphis, TN: National Cotton Council.

Tabashnik, B. E. 1990 Modelling and evaluation of resistance management tactics. In *Pesticide resistance in arthropods* (ed. R. T. Roush & B. E. Tabashnik), pp. 153–182. New York: Chapman & Hall.

Uchida, M., Asai, T. & Sugimoto, T. 1985 Inhibition of cuticle deposition and chiton synthesis by a new insect growth regulator, buprofezin, in *Nilaparvata lugens* Stal. *Pestic. Biochem. Physiol.* **27**, 71–75.

White, M. J. D. 1973 *Animal cytology and evolution*, 3rd edn. Cambridge University Press.

Williams, L., Dennehy, T. J. & Palumbo, J. C. 1997 Defining the risk of resistance to imidacloprid in Arizona populations of whitefly. In *Proceedings of the Beltwide Cotton Production Research Conference*, pp. 1242–1245. Memphis, TN: National Cotton Council.

Wolfenbarger, D. A. & Riley, D. G. 1994 Toxicity of mixtures of insecticides and insecticides alone against B-strain sweet-potato whitefly. In *Proceedings of the 1994 Beltwide Cotton Production Research Conference*, pp. 1214–1216. Memphis, TN: National Cotton Council.

Can anything be done to maintain the effectiveness of pyrethroid-impregnated bednets against malaria vectors?

C. F. Curtis, J. E. Miller, M. Hassan Hodjati, J. H. Kolaczinski and I. Kasumba

London School of Hygiene & Tropical Medicine, Keppel Street, London WC1E 7HT, UK

Pyrethroid-treated bednets are the most promising available method of controlling malaria in the tropical world. Every effort should be made to find methods of responding to, or preventing, the emergence of pyrethroid resistance in the *Anopheles* vectors. Some cases of such resistance are known, notably in *An. gambiae* in West Africa where the *kdr* type of resistance has been selected, probably because of the use of pyrethroids on cotton. Because pyrethroids are irritant to mosquitoes, laboratory studies on the impact of, and selection for, resistance need to be conducted with free-flying mosquitoes in conditions that are as realistic as possible. Such studies are beginning to suggest that, although there is cross-resistance to all pyrethroids, some treatments are less likely to select for resistance than others are. Organophosphate, carbamate and phenyl pyrazole insecticides have been tested as alternative treatments for nets or curtains. Attempts have been made to mix an insect growth regulator and a pyrethroid on netting to sterilize pyrethroid-resistant mosquitoes that are not killed after contact with the netting. There seems to be no easy solution to the problem of pyrethroid resistance management, but further research is urgently needed.

Keywords: *Anopheles*; malaria; pyrethroid resistance; insecticide-treated bednet; resistance management; organophosphates

1. INTRODUCTION

Malaria is by far the most important vector-borne disease, causing an estimated 300–500 million cases and 1.4–2.6 million deaths per year, 80–90% of them in Africa (WHO 1995). Pyrethroid-treated bednets have been shown in recent trials to have an important impact on (i) cases of malaria in China and India with low to moderate amounts of malaria transmission (e.g. Cheng *et al.* 1995; Jana-Kara *et al.* 1995), (ii) incidence of infection and/or prevalence of anaemia in areas of tropical Africa with moderate to intense malaria transmission (Stich *et al.* 1994; Snow *et al.* 1996; Curtis *et al.* 1998), (iii) hospital admissions in Kenya with severe malaria (Nevill *et al.* 1996), and (iv) all-cause child deaths in several parts of Africa (Alonso *et al.* 1993; D'Alessandro *et al.* 1995; Binka *et al.* 1996; Nevill *et al.* 1996).

In comparative trials with untreated nets, treated nets reduced malaria much more effectively (e.g. Jana-Kara *et al.* 1995) because the insecticide deposit reduces the chances of mosquitoes entering or biting through the net. Furthermore, where many people in a community use treated nets, these act as baited traps and kill a large proportion of the local mosquitoes before they can reach the age at which malaria parasites have reached maturity, thus reducing the malaria risk for the whole community.

Treated nets are more affordable and acceptable than house spraying (Kere & Kere 1992; Curtis *et al.* 1998), and it is extremely important that the early promise of the treated-net method is sustained. There are questions about long-term funding in very poor countries (Lengeler

et al. 1996) and about whether the reduction of vector populations without eradication will leave vulnerable those human populations that have not acquired their normal anti-malaria immunity (Snow *et al.* 1997). However, the subject of this paper is the threat to sustainability of this method from pyrethroid resistance in the *Anopheles* vectors, which might blunt the effectiveness of the method to the low and generally cost-ineffective levels now seen with untreated nets.

2. OCCURRENCE OF PYRETHROID RESISTANCE IN *ANOPHELES* SPECIES

In several dipteran species, resistance to knockdown by DDT (*kdr*) confers positive cross-resistance to pyrethroids (e.g. Farnham 1973; Prasittisuk & Busvine 1977), and it was feared that already existing resistance to DDT in anophelines would make them preadapted to resisting pyrethroids. However, several cases of DDT resistance were found to be due to metabolic mechanisms, which would not be expected to affect susceptibility to pyrethroids (Hemingway *et al.* 1985), and DDT-resistant *An. gambiae* from Zanzibar showed no cross-resistance to permethrin (R. T. Rwegoshora, unpublished data). However, by 1988, resistance to pyrethroids in several anopheline species had already been reported to the World Health Organization (WHO) (see tabulation by Malcolm (1988)).

Table 1 summarizes more recent published reports where pyrethroid resistance has apparently arisen as a result of selection by factors other than impregnated

Table 1. *Reports up until 1998 of cases of pyrethroid resistance in* Anopheles *spp. selected by factors other than use of impregnated bednets*

Anopheles spp.	country	reference(s)
albimanus	Guatemala	Beach *et al.* (1989)
sacharovi	Iraq	Akiyama (1996)
stephensi	India	Chakravorthy & Kalayasundaraman (1992)
stephensi	Dubai, UAE	Ladonni & Townson (1998), Vatandoost *et al.* (1998)
gambiae s.s.	Côte d'Ivoire and Burkina Faso	Elissa *et al.* (1993), Darriet *et al.* (1997), Martinez-Torres *et al.* (1998)

Table 2. *Results of tests for whether pyrethroid resistance had been selected by several years' use of impregnated bednets, as compared with where there had been no such use*

(Results from China and Tanzania are given in terms of median time for knockdown during exposure to nets treated with deltamethrin ($25 \, \mathrm{mg \, m^{-2}}$) or permethrin ($500 \, \mathrm{mg \, m^{-2}}$). The Kenyan results are presented as median time for mortality after timed exposures to paper impregnated with 0.25% permethrin.)

country	Anopheles spp.	KT_{50} or LT_{50} (min)				reference
China	years of net use	0	1	6	7	Kang *et al.* (1995)
(Hubei and Sichuan)	sinensis	11.0	11.6	11.3	8.7	
	anthropophagus	—	—	—	9.0	
Tanzania	years of net use	0		8		Curtis (1996)
(Tanga)	gambiae	10.1		12.2		
	funestus	7.2		7.3		
Kenya	years of net use	0	1	2	3	Vulule *et al.* (1996)
(Kisumu)	gambiae	13	31	28	36	

bednets. Resistance in *An. albimanus* in Guatemala was considered by Beach *et al.* (1989) to be connected with earlier selection by organophosphates (OPs). Earlier reports to WHO of resistance of *An. sacharovi* in Turkey and Syria have been extended by Akiyama (1996) to northern Iraq, but full susceptibility was found in anophelines from Baghdad and Basra. *An. stephensi* from Dubai (where much insecticide is used against nuisance insects) showed resistance to pyrethroids as well as to other insecticides (Ladonni & Townson 1998). The latter stock has been selected with permethrin in the laboratory to give a much higher level of resistance to a range of pyrethroids (as well as the 'near pyrethroid' etofenprox) and was used in the experiments with free-flying mosquitoes described below.

There have been several reports from Côte d'Ivoire and Burkina Faso of resistance in *An. gambiae s.s.* This is of the *kdr* type and involves cross-resistance to DDT, various pyrethroids and etofenprox (Elissa *et al.* 1993; Darriet *et al.* 1997; Martinez-Torres *et al.* 1998). The resistance gene can be detected by a method using the polymerase chain reaction and is thought to have been selected by pyrethroids used on cotton. Data will soon be published on laboratory and field tests of the effect of pyrethroid-impregnated netting on mosquitoes carrying this resistance gene (P. Guillet, personal communication).

Tests have been made at several sites in Sichuan and Hubei Provinces in China and in the village of Mng'aza in the Tanga Region of Tanzania. At the time of the tests, treated nets had been used in these places for 6–8 years and, in each country, comparisons were made with places where there had been no use of treated nets. As shown in table 2, no evidence of higher tolerance was found in the *Anopheles* spp. from the treated areas (Kang *et al.* 1995; Curtis 1996). In China the same testing method showed that strong resistance was being progressively selected during the time that a laboratory strain of *Culex quinquefasciatus* was maintained in a pyrethroid factory, presumably because of pollution of its environment by pyrethroid dust.

In Tanzania, recent tests (J. Myamba and C. Curtis, unpublished data) have included those in which larvae were collected close to villages, with or without long-term use of treated nets; the larvae were reared and the adults tested on the day of emergence. Such tests showed no evidence for resistance and, by using a standardized adult age for testing, avoided possible complications from the decline with age of pyrethroid tolerance which was found in both *An. stephensi* and *An. gambiae* (Hodjati & Curtis 1996). In addition, this avoided possible influences on the test results from any induction of tolerance that might arise by sub-lethal previous exposure of adult mosquitoes living in villages where many treated nets are in use.

As indicated above, a reduction in mean age of the local mosquito population is one of the achievable aims of community-wide introduction of treated nets. We would be interested to see age-standardized resistance tests in western Kenya where Vulule *et al.* (1994, 1996) reported a rise in permethrin tolerance of *An. gambiae* collected in the field as adults in four villages after one year of use of treated nets or curtains (table 2). There was no further rise after two more years' use. Such a pattern seems more

Table 3. *Numerical example to show the advantage, in a population, only part of which is exposed to an insecticide, if a dose can be used which kills all exposed heterozygotes (RS) for a resistance gene*

(Note that parents of the next generation are the sum of those not exposed plus those that survive exposure. Note that the rise in the frequency of R is far greater in case (ii) than in case (i).)

	RR	RS	SS	total	frequency of R
population at Hardy–Weinberg equilibrium	10	19 980	9 980 010	10 000 000	0.1%
10% unexposed to insecticide	1	1998	998 001	1 000 000	
(i) survivors of 90% exposure to dose that kills					
all RS and SS:	9	0	0	—	—
then parents of next generation:	10	1998	998 001	1 000 009	0.101%
(ii) survivors of 90% exposure to dose that kills					
75% of SS and 50% of RS:	9	8991	2 245 502	—	—
then parents of next generation:	10	10 989	3 243 503	3 254 502	0.169%

consistent with a life-shortening or other phenotypic effect of extensive use of pyrethroids in the area rather than with selection for resistance genes, which, once started, might be expected to proceed progressively so long as pyrethroid exposure continued. However, there is no doubt that resistance genes were present in the population, as artificial selection by Vulule *et al.* (1994) produced a strain with unequivocal resistance, although not as strong as that in the selected strain derived from *An. stephensi* of Dubai origin.

3. POSSIBLE METHODS OF USING PYRETHROIDS TO MINIMIZE THE RISK OF RESISTANCE IN *ANOPHELES*

If there is any truth in the general belief that low doses select for resistance, this could be due to a failure of low doses to kill resistance heterozygotes, i.e. rendering of the resistance genetically dominant or partly dominant at low doses, in contrast to recessiveness at high doses. From the Hardy–Weinberg ratio, new resistance genes, when still rare, would almost entirely be heterozygous until they have been selected to a moderate frequency. Thus, it is the heterozygous response that is crucial in the field because if it is only the very small number of resistance homozygotes that survive exposure, these will almost certainly be far outnumbered by susceptibles that avoid exposure, e.g. in houses without nets or in feeding and resting places other than dwelling houses. Thus, selection favouring resistance homozygotes alone would raise the frequency of the resistance gene only extremely slowly in the important early stages of build up of resistance (table 3).

In determining whether heterozygotes for resistance genes are less easily killed than susceptible homozygotes, it is important that the tests are made in conditions that are as realistic as possible. This is especially so with pyrethroids, which have an irritant effect on insects. We attempted laboratory simulations of the situation in a tropical bedroom provided with a treated net by sitting under such a net with an arm against it and releasing female mosquitoes to fly freely in a mosquito-proof laboratory room. We did not release males, because only females are attracted to blood feed and therefore we assumed that there would be little contact of males with the nets and so virtually no selection for resistance in that

sex. The females could be seen probing the net, and were often seen flying away and sometimes trying again. If they found the arm they might blood feed through the net but, with treated nets, many were found knocked down on the floor. We have so far been able to expose each of many replicate batches of mosquitoes for only half an hour, after which time the knocked-down and the still-active mosquitoes were collected and scored after 1 h and 24 h for delayed mortality or recovery from knockdown. Table 4a (based on Hodjati & Curtis (1997)) indicates that, with heterozygotes for the resistance of the selected Dubai strain of *An. stephensi*, there was a higher knock-down and mortality using a net treated with target doses of permethrin of 200 mg m^{-2} than with 500 mg m^{-2} (the actual doses were found by analysis to be somewhat less). On the lower dose there was 100% mortality after 24 h of both the heterozygotes and susceptible homozygotes, i.e. no selection for resistance can be expected. However, on the higher dose there was significantly lower mortality of the heterozygotes than the susceptibles. This apparently paradoxical lower mortality on the higher dose can be explained by the observed earlier take-off of mosquitoes irritated by netting with the higher dose, leading to shorter average exposure to the higher dose.

Thus, contrary to the accepted doctrine, a lower dose of this insecticide runs less risk of selecting for resistance than does a higher dose. We have for a long time advocated the use of a lower dose on the grounds of economy, in view of our data showing that its performance against susceptible wild mosquitoes is as good as that of the higher dose (Lines *et al.* 1987; Curtis *et al.* 1996). The argument for the use of a lower dose is further strengthened by the above data illustrating the relative risks of selecting for resistance.

Table 4b shows the preliminary data of I. Kasumba (unpublished) on various doses of lambdacyhalothrin found by analysis to range between 1 and 13 mg m^{-2}. The highest of these doses is approximately what is now commonly used for net treatment; markedly higher doses tend to make net users sneeze (Njunwa *et al.* 1991). At each of the doses tested, the resistance heterozygotes showed a knockdown rate that was distinctly lower than that of susceptibles, i.e. one could expect selection for resistance at each of these doses. We have been advocating this or similar alpha-cyano pyrethroids, in preference to permethrin, because the former are effective even after

Table 4. *Survival of susceptible (Beech stock) and heterozygous resistant (Beech × Dubai)* An. stephensi *after exposure to a human subject under a bednet with various alternative treatments (target dose and actual deposit found by gas liquid chromatography are indicated)*

(*a*) Data of Hodjati & Curtis (1997) with permethrin-treated nets and eight replicates for each genotype, each with 18–25 mosquitoes.

permethrin dose (mg m^{-2})		% survival 24 h after exposure		fitness of SS relative to RS (%)
target	actual	RS	SS	
0	0	100	100	100
200	147	0	0	—
500	470	23	5	22

(*b*) Data of I. Kasumba (unpublished) with various doses of lambdacyhalothrin. As indicated, some of the nets were washed and re-treated. Data are based on five replicate experiments with a total of *ca.* 90 mosquitoes tested on each net.

lambdacyhalothrin dose (mg m^{-2})		% still active 1 h after exposure		fitness of SS relative to RS (%)
target	actual	RS	SS	
0	0	93.6	85.1	90.9
3 (washed 1 ×)	1	75.8	42.6	56.2
3 (unwashed)	2	63.2	34.4	54.4
1 × 5 (washed 4 ×)	3	55.9	19.1	34.7
20 (unwashed)	13	28.5	12.5	43.9

washing nets and after treatment with much lower doses than are required where permethrin is used (Curtis *et al.* 1996). However, if there is confirmation of a reduced risk of selecting resistance when permethrin is used at a dose of 200 mg m^{-2}, then it would seem that we should make the latter the treatment of choice. This would be in line with the view urged by Roush (1989) that related compounds, to which there is cross-resistance, may nevertheless vary in the selective advantage that each gives to resistance genes, and the one giving the smallest selective advantage should be chosen. In our case, reverting to permethrin would require acceptance of a price in terms of higher treatment frequencies and costs in order to delay the onset of a resistance crisis. However, such a crisis would potentially be even more expensive in monetary terms and/or in lives lost from uncontrollable malaria.

4. POSSIBLE USE OF NON-PYRETHROIDS TO MANAGE PYRETHROID RESISTANCE IN *ANOPHELES*

Organophosphates (OPs) were tested on bednets by Brun & Sales (1976). Table 5 shows a summary of the more recent data of Miller *et al.* (1991), who compared bednets treated with pirimiphos methyl or pyrethroids in experimental huts. The OP performed as well as the pyrethroids in killing An. gambiae but not in protecting

Table 5. *Summary of data of Miller* et al. *(1991) on numbers of* An. gambiae *per night found to have entered, blood fed and died in experimental huts in The Gambia, in which human subjects slept under bednets with a standard pattern of holes to simulate a damaged net in domestic use*

(Means in the same column with different superscript letters differ at the 95% level of significance.)

net treatment	entered hut	blood fed	died
untreated	41.5[a]	11.1[a]	17.1[a]
permethrin	16.2[b]	4.5[b]	11.1[b]
lambda-cyhalothrin	33.8[a]	5.3[b]	30.5[c]
pirimiphos-methyl	48.5[a]	10.5[a]	48.0[d]

sleepers from being bitten, presumably because it lacks the irritancy of pyrethroids, which drives mosquitoes away after short contact or, with some formulations, even deters them from entering houses. The pirimiphos methyl formulation used caused stickiness of the net so that it attracted dirt, but we anticipate that the active ingredient could be reformulated to avoid this problem.

Different examples of resistance have been reported with either positive (e.g. Beach *et al.* 1989) or negative (Kurtak *et al.* 1987) cross-resistance between OPs and pyrethroids. The former situation would make OPs useless as a substitute for pyrethroids but the latter would make them an attractive means of actively driving pyrethroid resistance out of a population. The data of J. H. Kolaczinski (table 6) showed that, in *An. stephensi*, selection for resistance either to permethrin or malathion produced either no change or slight positive cross-resistance to the other compound. In *Cx quinquefasciatus*, however, selection for permethrin resistance had a negative effect on malathion tolerance. It remains to be seen how the *An. gambiae* populations in West Africa with pyrethroid resistance will respond to an organophosphate such as pirimiphos methyl on netting. Even if this has as good an insecticidal effect as shown in table 5 for a susceptible West African *An. gambiae*, and even though the toxicological profile of this compound is reassuring (Williams & White 1994), we anticipate some prejudice against the large-scale introduction of any OP for bedroom use in view of press coverage about the harmful effects of OP sheep dips to farmers and of OPs as a possible cause of the Gulf War syndrome.

The carbamate bendiocarb is considered safe enough for use on curtains but not on bednets. Tests against wild Tanzanian mosquitoes entering experimental huts showed that performance of bendiocarb-treated curtains was as good as pyrethroid-treated curtains (Curtis *et al.* 1996), However, in houses in Sri Lanka, pyrethroid-treated curtains performed significantly better than bendiocarb-treated curtains against a filariasis vector population of *Cx quinquefasciatus* (Weerasooriya *et al.* 1996).

The phenyl pyrazole fipronil has been tested on netting in the laboratory against anophelines, including the pyrethroid-resistant Dubai strain of *An. stephensi*. The netting was insecticidal but the time taken to kill the mosquitoes was so much delayed, compared with the action of pyrethroids (R. Williams and J. Kolaczinski, unpublished data), that it seems unlikely that it would assist in the

Table 6. *LT_{50} values (min) for different strains of mosquitoes selected with papers impregnated with 0.25% permethrin or 5% malathion in WHO insecticide-resistance test kits*

(Numbers in bold typeface indicate levels of resistance achieved by laboratory selection.)

species	strain	LT_{50} to 0.25% permethrin	LT_{50} to 5% malathion	cross-resistance to non-selecting compound
An. stephensi	Beech (susc.)	10.5	21.9	—
	Dubai	**486.2**	67.7	positive
	St Mal	9.4	**685.7**	none
Cx quinquefasciatus	PelSS (susc.)	51.7	42.7	—
	Muheza	**235.3**	29.5	negative
	Quinq	22.9	174.6	—
	Quinq (selected 19 generations)	**195.1**	36.4	negative[a]

[a]Compared with parental Quinq strain.

Table 7. *Tests of effects of insect growth regulators, which were picked up by tarsal contact with netting, on fecundity/fertility of female* Anopheles stephensi

(a) Data of J. E. Miller (1994, unpublished) on the effect of pyriproxyfen on fecundity during contact with either a polyester bednet or an Olyset net which incorporates permethrin into polyethylene fibre. Numbers of mosquitoes in parentheses.

Anopheles stephensi strain	mean no. of eggs laid after use of a net untreated with pyriproxifen	pyriproxifen-treated net		
		pyriproxifen dose (g m^{-2})	net with permethrin?	mean no. of eggs laid
Beech (pyrethroid susceptible)	84.3 (15)	0.5	no	56.1 (26)
	77.9 (10)	1.0	no	47.2 (15)
Dubai (pyrethroid resistant)	78.6 (37)	1.0	no	51.2 (22)
	69.1 (32)	1.0	yes	51.8 (36)

(b) Data of J. H. Kolaczinski (unpublished) on mean numbers of eggs laid after feeding for 5 min through untreated netting and netting treated with 1% triflumuron (numbers of mosquitoes tested in parentheses). There was a significant difference between numbers of eggs laid but not between % hatch.

Anopheles gambiae strain	untreated net	triflumuron-treated net
Kwale (pyrethroid susceptible)	75.4 (39)	53.1 (23)

personal protection of sleepers under torn nets in the way that pyrethroids do.

To sustain the control of the *Simulium* vectors, the Onchocerciasis Control Programme in West Africa successfully switches between different non-residual larvicides in rivers in response to detection of resistance to the preferred compound, temephos, and to seasonal changes in flow rate which limit the acceptability of some compounds to certain seasons (Hougard *et al.* 1993). This situation, where all residues are rapidly swept away, contrasts with a bednet treatment operation in which any attempt to switch insecticide when resistance was detected, or to operate a pre-planned policy of rotation, would be complicated by the persistence for a year or more of residues of insecticides applied earlier. Thus, mixtures of decaying and freshly applied residues with unpredictable selective effects would exist for long periods. In fact, however, mixtures of appropriate dosages of unrelated compounds may have better prospects of managing resistance effectively than do rotations (Mani

1985; Curtis *et al.* 1993; Barnes *et al.* 1995). Carefully chosen mixtures of several antibiotics (with directly observed consumption of the drugs to ensure patient compliance and meticulous hygienic practices) are recognized as the only way to control multi-drug resistance in outbreaks of tuberculosis, leprosy and other bacterial diseases.

We have been trying for several years to devise an appropriate mixture to use for bednet impregnation. The principal underlying the use of a mixture is that, so long as the resistance genes to each component are independent, rare and at linkage equilibrium, almost all resistant individuals would be only resistant to one of the components and would be killed by the other component provided that it was at the required dosage. As with the homozygotes in table 3, double-resistant combinations are assumed to be very rare compared with the proportion of the population that can be realistically expected to escape any selection. Among the ways that these assumptions can break down would be the occurrence of a mutant conferring positive cross-resistance. It seems unlikely that one could exclude this possibility for a pyrethroid–OP mixture nor, perhaps, for a pyrethroid–carbamate mixture.

As an alternative, we have tried to exploit the reported female-sterilizing properties of some insect growth regulators (IGRs) (Howard & Wall 1995). The intention is that one would apply a pyrethroid–IGR mixture to the nets in a community so that any pyrethroid-resistant mosquito, which made prolonged contact with the net and avoided being killed, would have simultaneously picked up enough of the IGR so as to be sterilized and thus unable to pass on its pyrethroid resistance genes.

Table 7*a* summarizes the data of Miller (1994, unpublished) demonstrating a considerable reduction in fecundity in pyrethroid-susceptible and pyrethroid-resistant *An. stephensi* that had attempted to feed on human subjects through nets carrying residues of the IGR pyriproxyfen. However, the infecundity was not complete even at a pyriproxyfen dosage that would be unaffordably high. More recently, J. H. Kolaczinski (unpublished data) has obtained similar results with the IGR triflumuron (table 7*b*).

No doubt this does not exhaust the possibilities of the use of IGR–pyrethroid mixtures for resistance and we understand that J. Bisset and M. Rodriguez in Cuba, with the encouragement of the WHO Tropical Diseases Research Programme, are embarking on a further quest for an effective IGR–pyrethroid mixture for use against Latin American anophelines. We wish them luck, as finding an effective means of preventing pyrethroid resistance in malaria vectors seems to us the most important practical current task in medical entomology.

We receive or have received financial support as follows: C.F.C., UK Medical Research Council; J.E.M., Zeneca and Sumitomo; M.H.H., Iranian Ministry of Health; J.H.K., Bayer and Rhône Poulenc; I.K., ODA and Hugh Pilkington Trust.

REFERENCES

Akiyama, J. 1996 Report to WHO East Mediterranean Regional Office, Alexandria. Alexandria: WHO.

Alonso, P. (and 10 others) 1993 A malaria control trial using insecticide-treated bed nets and targeted chemoprophylaxis in a rural area of The Gambia. *Trans. R. Soc. Trop. Med. Hyg.* **87** (Suppl. 2), 37–44.

Barnes, E. H., Dobson, R. J. & Barger, I. A. 1995 Worm control and antihelmintic resistance: adventures with a model. *Parasitol. Today* **11**, 56–63.

Beach, R. F., Cordón-Rosales, C. & Brogdon, W. G. 1989 Detoxifying esterases may limit the use of pyrethroids for malaria vector control in the Americas. *Parasitol. Today* **5**, 326–327.

Binka, F., Kubaje, A., Adjuik, M., Williams, L., Lengeler, C., Maude, G. H., Armah, G. E., Kajihara, B., Adiamah, J. H. & Smith, P. G. 1996 Impact of impregnated bednets on child mortality in Kassena-Nankana, Ghana: a randomized controlled trial. *Trop. Med. Int. Hlth* **1**, 147–154.

Brun, L. O. & Sales, S. 1976 Stage IV evaluation of four organophosphates. WHO mimeographed document. WHO/VBC/76.630. Geneva: WHO.

Chakravorthy, B. C. & Kalyasundaraman, M. 1992 Selection of permethrin resistance in the malaria vector *Anopheles stephensi*. *Ind. J. Malariol.* **29**, 161–165.

Cheng, H., Yang, W., Kang, W. & Liu, C. 1995 Large-scale spraying of bednets to control mosquito vectors and malaria in Sichuan, China. *Bull. Wld Hlth Org.* **73**, 321–328.

Curtis, C. F. 1996 Detection and management of pyrethroid resistance in relation to the use of impregnated bednets against malaria vectors. In *2nd Int. Conf. Ins. Pests Urban Envir.* (ed. K. D. Wildey), pp. 381–384. Edinburgh.

Curtis, C. F., Hill, N. & Kasim, S. 1993 Are there effective resistance management strategies for vectors of human disease? *Biol. J. Linn. Soc.* **48**, 3–18.

Curtis, C. F., Myamba, J. & Wilkes, T. 1996 Comparison of different insecticides and fabrics for anti-mosquito bednets and curtains. *Med. Vet. Entomol.* **10**, 1–11.

Curtis, C. F., Maxwell, C. A., Finch, R. & Njunwa, K. J. 1998 Comparison of use of a pyrethroid for house spraying or bednet treatment against Tanzanian malaria vectors. *Trop. Med. Int. Hlth* **3**. (In the press.)

D'Alessandro, U., Olaleye, B. O., McGuire, W., Langerock, P., Aikins, M. K., Thomson, M., Bennett, S., Cham, M. K. & Greenwood, B. M. 1995 Reduction of mortality and morbidity from malaria in Gambian children following introduction of a National Insecticide Impregnated Bednet Programme. *Lancet* **345**, 479–483.

Darriet, F., Guillet, P., Chandre, F., N'Guessan, R., Doannio, J. M. C., Rivière, F. & Carnevale, P. 1997 Présence et évolution de la résistance aux pyréthrinoides et au DDT chez deux populations d'*Anopheles gambiae s.s.* d'Afrique de l'ouest. WHO mimeographed document. Geneva: WHO/CTD/VBC/97.1001.

Elissa, N., Mouchet, J., Rivière, F., Meunier, J.-Y. & Yao, K. 1993 Resistance of *Anopheles gambiae s.s.* to pyrethroids in Côte d'Ivoire. *Ann. Soc. Belge Méd. Trop.* **73**, 291–294.

Farnham, A. W. 1973 Genetics of resistance of pyrethroid-selected houseflies, *Musca domestica*. *Pestic. Sci.* **4**, 513–520.

Hemingway, J., Malcolm, C. A., Kissoon, K. E., Boddington, R. G., Curtis, C. F. & Hill, N. 1985 The biochemistry of insecticide resistance in *Anopheles sacharovi*: comparative study with a range of insecticide susceptible and resistant *Anopheles* and *Culex* species. *Pestic. Biochem. Physiol.* **24**, 68–76.

Hodjati, M. H. & Curtis, C. F. 1996 Pyrethroid resistance in *Anopheles* is age dependent. *Ann. Trop. Med. Parasitol.* **90**, 438. (Abstract.)

Hodjati, M. H. & Curtis, C. F. 1997 Dosage differential effects of permethrin impregnated into bednets on pyrethroid resistant and susceptible genotypes of the mosquito *Anopheles stephensi*. *Med. Vet. Entomol.* **11**, 368–372.

Hougard, J.-M., Poudiougo, P., Guillet, P., Back, C., Akpoboua, I. K. B. & Quillévéré, D. 1993 Criteria for the selection of larvicides by the Onchocerciasis Control Programme in West Africa. *Ann. Trop. Med. Parasitol.* **87**, 435–442.

Howard, J. & Wall, R. 1996 Autosterilization of the house fly *Musca domestica* using the chitin synthesis inhibitor triflumuron on sugar-baited targets. *Med. Vet. Entomol.* **10**, 97–100.

Jana-Kara, B. R. J., Wajihullah, Shahi, B., Vas Dev, Curtis, C. F. & Sharma, V. P. 1995 Deltamethrin impregnated bednets against *Anopheles minimus* transmitted malaria in Assam, India. *J. Trop. Med. Hyg.* **98**, 73–83.

Kang, W., Gao, B., Jiang, H., Wang, H., Yu, T., Yu, P., Xu, B. & Curtis, C. F. 1995 Tests for possible effects of selection by domestic pyrethroids for resistance in culicine and anopheline mosquitoes in Sichuan and Hubei, China. *Ann. Trop. Med. Parasitol.* **89**, 677–684.

Kere, J. F. & Kere, N. K. 1992 Bed-nets or spraying? Cost analyses of malaria control in the Solomon Islands. *Hlth Policy Planning* **7**, 382–386.

Kurtak, D., Meyer, R., Ocran, M., Ouédrago, M., Renaud, P., Sawadogo, R. O. & Télé, B. 1987 Management of insecticide resistance in control of the *Simulium damnosum* complex by the

Onchocerciasis Control Programme, West Africa: potential use of negative correlation between organophosphate resistance and pyrethroid susceptibility. *Med. Vet. Entomol.* **1**, 137–146.

Ladonni, H. & Townson, H. 1998 A major gene conferring permethrin resistance in the larvae of the malaria vector *Anopheles stephensi. Bull. Entomol. Res.* (In the press.)

Lengeler, C., Cattani, J. & de Savigny, D. (eds) 1996 *Net gain: a new method of preventing malaria deaths.* Ottawa: International Development Research Centre; and Geneva: WHO.

Lines, J. D., Myamba, J. & Curtis, C. F. 1987 Experimental hut trials of permethrin-impregnated mosquito nets and eave curtains against malaria vectors in Tanzania. *Med. Vet. Entomol.* **1**, 37–51.

Malcolm, C. 1988 Current status of pyrethroid resistance in anophelines. *Parasitol. Today* **4**, S13–S15.

Mani, G. S. 1985 Evolution of resistance in the presence of two insecticides. *Genetics* **109**, 761–783.

Martinez-Torres, D., Chandre, F., Williamson, M. S., Darriet, F., Bergé, J. B., Devonshire, A. L., Guillet, P., Pasteur, N. & Pauron, D. 1998 Molecular characterization of pyrethroid knockdown resistance (*kdr*) in the major malaria vector *Anopheles gambiae* s.s. *Insect Molec. Biol.* **7**, 179–184.

Miller, J. E. 1994 Can pyriproxyfen (an insect growth regulator) be used to prevent permethrin resistance by impregnated bednets. *Trans. R. Soc. Trop. Med. Hyg.* **88**, 281. (Abstract.)

Miller, J. E., Lindsay, S. W. & Armstrong, J. R. M. 1991 Experimental hut trials of bednets impregnated with synthetic pyrethroid or organophosphate insecticide for mosquito control in The Gambia. *Med. Vet. Entomol.* **5**, 465–476.

Nevill, C., Some, E. S., Mung'ala, V. O., Mustemi, W., New, L., Marsh, K., Lengeler, C. & Snow, R. W. 1996 Insecticide-treated bed nets reduce mortality and severe morbidity from malaria among children on the Kenyan coast. *Trop. Med. Int. Hlth* **1**, 139–146.

Njunwa, K. J., Lines, J. D., Magesa, S. M., Mnzava, A. E. P., Wilkes, T. J., Alilio, M., Kivumbi, K. & Curtis, C. F. 1991 Trial of pyrethroid impregnated bednets in an area of Tanzania holoendemic for malaria. 1. Operational methods and acceptability. *Acta Trop.* **49**, 87–96.

Prasittisuk, C. & Busvine, J. R. 1977 DDT-resistant mosquito strains with cross-resistance to pyrethroids. *Pestic. Sci.* **8**, 527–534.

Roush, R. 1989 Designing resistance management programs: how can you choose? *Pestic. Sci.* **26**, 423–441.

Snow, R. W., Molyneux, C. S., Warn, P. A., Omumbo, J., Nevill, C. G., Gupta, S. & Marsh, K. 1996 Infant parasite rates as a measure of exposure to *Plasmodium falciparum* during a randomized controlled trial of insecticide-treated bed nets on the Kenyan coast. *Am. J. Trop. Med. Hyg.* **55**, 144–149.

Snow, R. W., Omumbo, J. A. & Lowe, B. 1997 Relation between severe malaria morbidity in children and level of *Plasmodium falciparum* transmission in Africa. *Lancet* **349**, 1650–1654.

Stich, A. H. R., Maxwell, C. A., Haji, A. A., Haji, D. M., Machano, A. Y., Mussa, J. K., Matteeli, A., Haji, H. & Curtis, C. F. 1994 Insecticide-impregnated bed nets reduce malaria transmission in rural Zanzibar. *Trans. R. Soc. Med. Hyg.* **88**, 150–154.

Vatandoost, H., McCaffery, A. & Townson, H. 1998 An electrophysiological investigation of target-site insensitivity in permethrin-resistant and permethrin-susceptible strains of *Anopheles stephensi. Bull. Entomol. Res.* (In the press.)

Vulule, J., Beach, J. M., Atieli, F. K., Roberts, J. M., Mount, D. L. & Mwangi, R. W. 1994 Reduced susceptibility of *Anopheles gambiae* to permethrin associated with the use of permethrin-impregnated bednets and curtains in Kenya. *Med. Vet. Entomol.* **8**, 71–75.

Vulule, J., Beach, J. M., Atieli, F. K., Mount, D. L., Roberts, J. M. & Mwangi, R. W. 1996 Long term use of permethrin-impregnated nets does not increase *Anopheles gambiae* tolerance. *Med. Vet. Entomol.* **10**, 71–79.

Weerasooriya, M. V., Munasinghe, C. S., Mudalige, M. P. S., Curtis, C. F. & Samarawickrema, W. A. 1996 Comparative efficacy of house curtains impregnated with permethrin, lambdacyhalothrin or bendiocarb against the vectors of filariasis. *Trans. R. Soc. Trop. Med. Hyg.* **90**, 103–104.

Williams, N. & White, G. B. 1994 "Actellic[R]" (pirmiphos-methyl) for the control of *Aedes* vectors in dengue control programmes. In *First Int. Cong. Parasitol. Trop. Med.*, Kuala Lumpur, pp. 113–117.

World Health Organization 1995 Vector control for malaria and other mosquito-borne diseases. *WHO Tech. Rep. Ser.* **857**, 2.

Two-toxin strategies for management of insecticidal transgenic crops: can pyramiding succeed where pesticide mixtures have not?

R. T. Roush

Department of Crop Protection, University of Adelaide, Waite Campus, PMB1, Glen Osmond, South Australia 5064, Australia
(*rroush@waite.adelaide.edu.au*)

Transgenic insect-resistant crops that express toxins from *Bacillus thuringiensis* (Bt) offer significant advantages to pest management, but are at risk of losing these advantages to the evolution of resistance in the targeted insect pests. All commercially available cultivars of these crops carry only a single Bt gene, and are particularly at risk where the targeted insect pests are not highly sensitive to the Bt toxin used. Under such circumstances, the most prudent method of avoiding resistance is to ensure that a large proportion of the pest population develops on non-transgenic 'refuge' hosts, generally of the crop itself. This has generated recommendations that 20% or more of the cotton and maize in any given area should be non-transgenic. This may be costly in terms of yields and may encourage further reliance on and resistance to pesticides. The use of two or more toxins in the same variety (pyramiding) can reduce the amount of refuge required to delay resistance for an extended period. Cross-resistance among the toxins appears to have been overestimated as a potential risk to the use of pyramids (and pesticide mixtures) because cross-resistance is at least as important when toxicants are used independently. Far more critical is that there should be nearly 100% mortality of susceptible insects on the transgenic crops. The past failures of pesticide mixtures to manage resistance provide important lessons for the most efficacious deployment of multiple toxins in transgenic crops.

Keywords: *Bacillus thuringiensis*; insect resistance; *Helicoverpa*; cotton; corn borer; maize

1. INTRODUCTION

For more than 50 years, the breeding of crop cultivars that suffered reduced losses to insects has played a major role in pest management research (Painter 1951). However, in spite of significant successes, particularly against pests that attack crops in a single plant genus or family (Maxwell & Jennings 1980), progress on classical 'host-plant resistance' is often slow and the overall impact on pest management has been limited. Molecular genetic engineering offers the potential to introduce new insect resistance traits across species barriers, as first demonstrated in parallel studies on tobacco in 1987 using genes from the bacterium *Bacillus thuringiensis* (Bt) and cowpea (Schuler *et al.* 1998).

The most successful of these transgenic crops, and the only ones that have been commercially released, have been those using the crystal protein endotoxin (*cry*) genes from Bt, especially those that produce Cry 1A toxins. After their commercial introduction in 1996, transgenic Cry 1A cotton and maize were planted on one million and three million hectares ($1\,ha = 10^4\,m^2$), respectively, in 1997 (Tabashnik *et al.*, this issue). Transgenic potatoes producing a Cry 3A toxin for control of the Colorado potato beetle (*Leptinotarsa decemlineata*) are very effective (Feldman & Stone 1997), but were grown on only 10 000 hectares in 1997 (Tabashnik *et al.*, this issue), largely due to strong competition from new insecticides (especially imidacloprid) that control a wider range of pests.

Bt transgenic crops can significantly reduce the use of insecticides, increasing the abundance of non-target and beneficial species in crops (e.g. Fitt *et al.* 1994), and reducing the need for insecticidal sprays even for pests not targeted by the transgenics (e.g. Feldman & Stone 1997). In the case of cotton, actual reductions in use have been in the range of 50–60% (Roush & Shelton 1997). The reduction of insecticide sprays in crops may have particular significance in the tropics. There is considerable evidence that the use of such sprays is a major selective force for insecticide resistance in pests of medical importance, particularly mosquitoes (Georghiou 1990), which although not targeted by agricultural sprays are nonetheless in the fields at the times the sprays are made. Key factors among these cross-ecosystem problems are the impacts of agricultural use of pyrethroid insecticides that affect the control of *Anopheles* mosquitoes by insecticide-impregnated bednets. The malaria transmitted by these mosquitoes causes more than one million deaths per year, mostly of African children (Curtis *et al.*, this issue). At least some of this pyrethroid use is on crops currently targeted for Bt transgenic technology, including cotton.

Therefore Bt transgenic crops offer a number of benefits for the environment and human health, but these benefits are at risk due to the potential for resistance (Tabashnik *et al.*, this issue). Bt sprays have been used for decades, but they are of such short persistence when applied in the field that they generally have poor efficacy against most insects and therefore have had limited use (Roush 1994). One exception is the diamondback moth, *Plutella xylostella*, which was intrinsically quite susceptible to Bt toxins, and suffered considerable Bt use when resistance had evolved to other insecticides. The diamondback moth has now evolved resistance to Bt sprays in many if not most areas of the tropics (Tabashnik 1994a; Perez & Shelton 1997; Tabashnik *et al.*, this issue). Bt transgenic crops significantly increase the efficacy of the Cry toxins compared to Bt sprays, and therefore the potential for resistance, perhaps primarily because transgenic crops are more likely to be used than Bt sprays (Roush 1994).

Owing to the current commercial significance of Bt crops, the urgency of resistance management plans for their deployment, and the depth of public concern about the future of Bt, this paper will concentrate on Bt transgenic crops. However, the general principles applied to Bt crops will also be relevant to other kinds of insecticidal transgenic crops.

2. RESISTANCE MANAGEMENT TACTICS FOR BT CROPS

Before developing the main topic of this paper—an exploration of the use of toxin stacking or pyramiding—it will help to put pyramiding in context with an overview of tactics that have been proposed for resistance management for Bt transgenic crops. There are at least eight possible types of tactics to slow selection by transgenic plants, some of which are mutually exclusive: (i) express toxin genes only moderately strongly, so that not all susceptible individuals are killed; (ii) modify the expression of the genes in each plant, such that they are expressed only as needed to protect the crop through tissue-specific, temporal-specific or inducible promoters; (iii) express the toxins to as high as is agronomically acceptable; (iv) deploy different toxins individually in different varieties, simultaneously; (v) deploy toxins sequentially, i.e. reserve toxins until previous ones fail; (vi) deploy plants with a mixture of toxins; (vii) leave non-transgenic crop and non-crop host-plants as 'refuges' for susceptible insects; and (viii) deploy the crops as part of an overall integrated pest management (IPM) programme that combines multiple tactics for control.

Owing to the lack of technological feasibility, poor efficacy of resistance management and impracticality of pest control, low and variable expression ((i) and (ii) from the list above) have been essentially abandoned as viable deployment options, at least for the near term, in favour of (iii) high expression (Roush 1996, 1997a; Gould 1998). The success of high expression depends critically on the use of (vii) a refuge (Roush 1994; Gould 1998). The need for refuges provides a strong incentive for integrating transgenic crops within a more general IPM programme (viii), in some cases potentially using transgenic crops as a means of regulating pest population growth to support other non-insecticidal tactics (Roush 1997a). As will be discussed in § 5, pyramiding or stacking multiple toxins in the same plants (vi) appears to be a much more effective strategy than deploying them individually ((iv) or (v)).

Currently, all cultivars of insecticidal transgenic crops that are available commercially carry only a single Bt gene. However, single-toxin crops are particularly at risk where the targeted insect pests are not highly sensitive to the Bt toxin used (Roush 1997b; Gould & Tabashnik 1998). As will be discussed in § 4, the most prudent method of avoiding resistance under such circumstances is to ensure that a large proportion of the pest population develops on non-transgenic 'refuge' hosts, generally of the crop itself. This has generated recommendations that 20% or more of cotton and maize should be non-transgenic in Australia (R. T. Roush, unpublished results) and North America (Gould & Tabashnik 1998; Andow & Hutchison 1998). Such high refuges may cost in terms of yields, encourage further reliance on pesticides, and exacerbate resistance to the pesticides used. Pyramids offer the potential for superior delays in resistance with smaller and more acceptable refuge sizes (Roush 1997b).

3. SIMULATION MODELLING

(a) *Underlying assumptions*

This paper will make extensive use of simple computer simulation models. It would be preferable to make decisions for resistance management on the basis of realistic experiments in the field, but these would take years and considerable expense (both in money and delays of the environmental and health benefits). Simulation models are our best tools in the foreseeable future.

Specialized versions of the same basic simulation model were developed for one or two toxins in the plants, to which the insect population could respond to by up to three resistance loci. Except as noted, both toxins used in a pyramided plant were assumed to be at risk for the development of resistance. Initially, two loci in the insects were specific to toxins A and B, respectively, but in later simulations (described in § 5f) a third locus was added that could confer resistance to both toxins. Except as noted, the initial frequency of resistance alleles at each locus was 10^{-3}, following the conclusions of Gould *et al.* (1997) for a cotton bollworm, *Heliothis virescens*. This assumption is not an endorsement of the estimate of Gould *et al.* (1997); to the contrary, their data seem inconsistent even with the predictions given within their paper for single-gene inheritance of the resistance studied. Nonetheless, because resistance management is more effective when the frequency of resistance is lower (Roush 1994, 1997a), an assumption of 10^{-3} should be conservative.

To be even more conservative, the models assumed that there are no fitness-costs to resistance. If fitness-costs do exist, pyramiding strategies will work even better compared to sequential deployment (Gould 1994, 1998), as can also be true for pesticide mixtures (Roush 1989). These simulations also assumed that resistant homozygotes were unaffected by the plants, and except as noted, that susceptible larvae always died when they fed on transgenic plants. Except as noted, the simulations all assumed that 10% of eggs in the population are laid on non-transgenic hosts. Implicitly, eggs laid on non-transgenic hosts survived equally well as resistant homozygous insects on transgenic hosts. Population growth was density independent. The models also assumed that selective toxin exposure occurred only for larvae, as would be true for lepidopteran pests on Bt transgenic crops. The

Cry 3A transgenic Bt potato plants mentioned above, control both adult and larval Colorado potato beetles (Feldman & Stone 1997). Single-locus models developed for selection on both adults and larvae of Colorado potato beetles (Roush 1996; Gould *et al.* 1994) essentially give the same results as discussed here. The extremely high sensitivity of potato beetles to Cry 3A toxin encourages optimism about resistance management for that species (Roush 1994).

Mating was assumed to be random throughout the populations (implying that transgenic and non-transgenic hosts are close enough to one another for moths to freely exchange) and, in the absence of selection, the frequencies of the genotypes are based on the Hardy–Weinberg expression (where p represents the frequency of the resistance allele (R) and q the susceptible allele (S), p^2 gives the frequency of RR homozygotes, $2pq$ for RS heterozygotes, and q^2 for SS homozygotes).

It is further assumed that larvae do not move between transgenic and non-transgenic hosts, i.e. that there are neither seed mixtures nor row mixtures of transgenic and non-transgenic hosts that allow interplant dispersal. Significant exchange of larvae between transgenic and non-transgenic plants, which has been observed in cotton (R. T. Roush and G. P. Fitt, unpublished results), maize (P. Davis and R. T. Roush, unpublished results), rice (Bennett *et al.* 1997) and broccoli (Shelton *et al.* 1998), can accelerate resistance in theory (Mallet & Porter 1992; Tabashnik 1994*b*; Roush 1996) and has done so in experiments with diamondback moth (Shelton *et al.* 1998). Seed mixtures can also allow significant damage to the crop, at least in cotton (R. T. Roush and G. P. Fitt, unpublished results).

(b) *Modelling format*

All versions of the simulation model used the same deterministic general format. Populations distribute their eggs in Hardy–Weinberg proportions at random across refuge and transgenic habitats, with larvae suffering mortality based on their genotypes. Where some percentage of the population escapes exposure to the toxins, the model simply sets that fraction of the population aside from a selection routine. In the absence of the third locus with cross-resistance, linkage between loci was followed through frequencies of all ten possible unique two-locus genotypes (including coupling and repulsion). Recombination (r) between the loci was adjusted from 0.01 to 0.5. Except as noted, the time until resistance evolves is measured as the number of (non-overlapping) generations until the frequency of the resistance allele exceeds 50%. This is a convenient measure of resistance, which is independent of assumptions on population growth. At the level of changes in frequency of the resistance allele, the models were checked against similar models (e.g. Mani 1985; Mallet & Porter 1992; Tabashnik 1994*b*) and gave the same results for the same parameter values. The model also tracks population density on both transgenic and non-transgenic host plants.

4. SINGLE-TOXIN BT TRANSGENIC CROPS

The mortality of heterozygous (RS) insects is the strongest influence on the rate of evolution of resistance

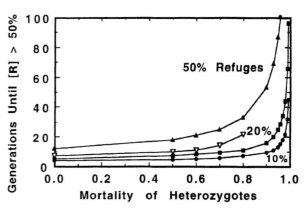

Figure 1. Effect of mortality of RS heterozygous larvae and proportion of eggs that developed on non-transgenic refuge hosts on the evolution of resistance. As discussed in the text, results are from a simulation model with the following assumptions: a single locus, random mating, no selective mortality of resistant homozygous larvae, that some fraction of the population escapes exposure (refuges of 10% (filled circles), 20% (filled squares and open triangles) or 50% (filled triangles)), and initial frequencies of resistance allele of 10^{-3}. For all of the filled symbols, it was assumed that all susceptible insects died when they fed on transgenic plants; the curve with the open symbols represents a case in which 10% of the susceptibles on transgenic plants survived when there was a 20% refuge.

for insecticidal crops that produce only a single Bt toxin, provided that there is a significant refuge to produce moths that can mate with resistant survivors and ensure that most of their offspring will be heterozygous or susceptible (Roush 1997*a*,*b*). When there is only one resistance locus in the insects and the mortality of heterozygous individuals exposed to transgenic crops exceeds 95%, resistance can be delayed for more than 40 generations even when resistance is initially as common as 10^{-3} and only 10% of the pest population develops on refuge hosts (figure 1). This is the basis of the 'high-expression' or 'high-kill' strategy for resistance management. However, when the mortality of heterozygotes is less than 90%, one needs a refuge of more than 20% to delay resistance for more than 20 generations, and greater than 10% even if the initial resistance frequency is 10^{-6} (Mallet & Porter 1992; Roush 1994, 1997*a*,*b*). Not even the survival of 10% of the susceptible individuals on transgenic plants can do much to delay resistance (figure 1). Where the survival of susceptibles does appear to cause a delay of resistance (i.e. at 80% mortality of RS in figure 1), it is only because resistance is effectively very recessive.

Although 100% mortality of Bt-resistant heterozygotes has been observed for *Heliothis virescens* (Gould *et al.* 1997) and the diamondback moth (Roush 1994; Metz *et al.* 1995), it seems unlikely that all species targeted by some or all current single-toxin Bt cultivars are as well controlled, especially *Helicoverpa* species on maize and cotton (Roush 1997*b*; Andow & Hutchison 1998; Gould 1998; Gould & Tabashnik 1998). In such cases, the only way to delay resistance for single-toxin plants is with very large refuges (figure 1). Pyramids offer a means to delay resistance with practically acceptable refuge sizes (Roush 1997*b*), as described next.

5. TWO TOXINS: PYRAMIDING

(a) *Options for two-toxin deployment*

Given two different insecticidal toxins that are believed not to share cross-resistance (i.e. there is no one mechanism in the pests that can confer resistance to both), there are three general choices as to how the toxins could be deployed: (i) individually but simultaneously in different varieties, i.e. as a mosaic (either as a seed mix within the same field or in neighbouring fields); (ii) sequentially (one after another in an evolutionary race with the pests); or (iii) they could be stacked in the same variety, i.e. pyramided (see § 2). Previous studies have shown that sequential deployment is always at least as good as, and often superior to, mosaics for insecticides and toxins (Roush 1989, 1997*a,b*). Given that the development of toxin cultivars is a commercial exercise, it seems unlikely that withholding resistance genes, especially when developed by competing companies, will be seen as consistent with a free market. However, because sequential deployment is a better option than mosaics, pyramids will be compared with sequential deployment in the following discussion.

In contrast to single-toxin 'high-kill' strategies, pyramiding relies on the idea that each toxin is used individually in a way that would kill all insects susceptible to that toxin, and in so doing, kills insects that are resistant to the companion toxin (Roush 1997*a*). This is 'redundant killing' in the sense that most of the population, which is susceptible to both toxins, is killed twice (Comins 1986; Gould 1986*a,b*). The extent to which individuals that are resistant to one toxin are killed by the other is central to the effectiveness of the pyramiding strategy.

(b) *Candidate toxins for pyramiding*

Ideally, the two toxins should be as unrelated as possible, such as a Bt toxin and a digestive inhibitor, to minimize the chance that a single gene in the pest species could confer resistance to both factors. As a practical matter, the only toxins that will likely be available for pyramiding in at least the next five years will be Bt Cry proteins. Although there is a wide range of alternative toxins under development, including inhibitors of digestive enzymes (proteinase and amylase inhibitors), lectins, chitinases and peroxidases (Carozzi & Koziel 1997; Schuler *et al.* 1998), few have yet proven to be effective against the same pests for which Bt toxins are being successfully deployed. For example, although an alpha-amylase inhibitor in seed-targeted expression is very effective for the control of pea weevils (Schroeder *et al.* 1995), no complementary Bt toxin is known. The more promising candidates that might complement Bt toxins include cholesterol oxidase against *Helicoverpa* and *Heliothis* species (Purcell 1997), and proteinase inhibitors against rice stem borers (Bennett *et al.* 1997). Although often more effective in tobacco than other crops, non-Bt toxins typically cause only a 30–80% delay in development rather than a practically significant increase in mortality (Carozzi & Koziel 1997; Schuler *et al.* 1998). Other newer candidates, such as *Helicoverpa* stunt virus (Schuler *et al.* 1998) and toxins from *Photorhabdus luminescens*, are probably at least five to ten years from commercial

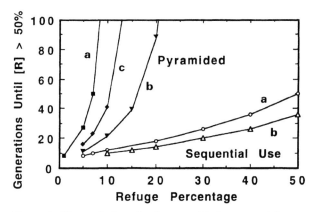

Figure 2. The evolution of resistance with the sequential deployment of two toxins compared with the use of the toxins jointly in a pyramided variety, for a range of percentages of the population in refuges. For comparison, it was assumed that there was 70% mortality of RS heterozygotes for each toxin (curve 'a'), 50% mortality of RS heterozygotes for each toxin (curve 'b'), or 50% mortality of RS for one toxin and 70% for the other (curve 'c'). The values shown for 'sequential' curves are twice the number of generations required for resistance to evolve to just one of the toxins when used alone, as if cultivars bearing one and then the other of the toxins were sequentially deployed.

release. More than ten years were required to develop Bt transgenic crops using toxin genes that were already well characterized by 1986.

Even for Bt toxins, probably no more than two with effectiveness against any given pest will be useful and available in the next five years. Given that decreased binding has proven to be a common mechanism of resistance, the appropriate toxins should at least have different binding sites. Further, they should not use toxins that have already shown significant levels of cross-resistance in strains observed to date, such as between Cry 1A, Cry 1F and Cry 1J (Tabashnik *et al.* 1997*a*, this issue). However, at least two good toxins seem to be available for most key pests, such as CrylA and Cry 9C for corn borers (Roush 1997*b*).

(c) *Refuge needs for one- and two-toxin plants without cross-resistance*

Even though individuals with resistance to two toxicants may be very rare initially, a refuge is still necessary for pyramiding to be effective in delaying resistance (Curtis 1985; Gould 1986*a,b*; lower left of figure 2, refuge less than 5%). However, when selection can respond only at separate toxin-specific loci in the insects, the refuge for a similar delay of resistance can be much smaller when toxins are pyramided than if they are sequentially deployed (Roush 1997*b*). As long as at least 50% of the heterozygotes are killed when they feed on transgenic plants, a pyramid with even a 10% refuge can delay resistance for longer than if the two toxins are sequentially deployed with a 30–40% refuge (figure 2, compare pyramid curves 'a' and 'b' with sequential curves 'a' and 'b'). If the control of heterozygotes is less than 50% for both toxins, essentially neither strategy will be effective for delaying resistance (e.g. see lower left of figure 1).

As with insecticide mixtures (Mani 1985), pyramids are most effective when at least one of the resistances is mostly recessive (Gould 1986a,b). If resistance to at least one toxin confers no more than 30% survival in a pyramid, even a 10% refuge can support a significant delay of resistance (figure 2, compare pyramid curves 'b' and 'c'). However, as with single toxins, it is still desirable for the expression of both toxins to be as high as is agronomically acceptable, both for the control of heterozygotes and, as will be seen in the next subsection, for the control of susceptible homozygotes.

Even though 100% of heterozygous *H. virescens* and diamondback moths have been killed in experiments on current Bt cultivars, current cultivars do not even kill 100% of susceptible larvae among *Helicoverpa* species. Based on extrapolations of diamondback moths (§ 5d), one might expect that plants that kill around 95% of susceptible homozygotes (in the range of *Helicoverpa* species on current Bt cotton cultivars) would likely result in 50–70% mortality of heterozygotes. However, a key feature of the pyramiding strategy is that only one of the types of heterozygotes needs to have such high mortality; pyramiding two or more toxins into a cultivar increases the chance that at least one will be especially favourable to resistance management.

Because pyramids can reduce the need for large refuges, they provide a way to use Bt genes in cotton without relying on maize and other crops as refuges for *Helicoverpa* species (Roush 1997b). This is especially important in light of the expected releases of Bt cotton and maize into the USA and Mexico, South America, Africa, China and India in the same regions with the same *Cry 1A* gene.

(d) *Effect of mortality of susceptible homozygotes on pyramids*

The success of pyramids in the absence of cross-resistance is less dependent on high mortalities of heterozygotes than a single toxin. Whereas the single toxin requires large refuges whenever mortalities of heterozygotes are less than 90% (figure 1), pyramids can still be effective with relatively small refuges of 10% even for heterozygous mortalities of 30–50% (figure 2). However, in contrast to plants with single toxins (figure 1), the survival of susceptible insects on transgenic plants has a major effect on the durability of pyramids (Roush 1994; figure 3). Whereas the high-kill strategy aims to control heterozygotes directly, the pyramiding strategy aims to do so by killing the individuals resistant to one toxin with a second toxin. The requirement to exceed 95% kill to achieve significant benefits still applies, but in this case, it is the mortality of susceptible homozygotes that matters (figure 3).

In considering what would be reasonable parameter values for these simulations, I used the example of Bt-resistant diamondback moths from Hawaii and Florida, for which concentration mortality data are provided by Tabashnik *et al.* (1992) and Tang *et al.* (1997). Resistance in these strains appears to be primarily due to a single major gene (Tabashnik *et al.* 1997b; Tang *et al.* 1997), so the F1 larvae used in bio-assays are presumed to be heterozygotes. In these populations, Bt concentrations that cause 99%, 95% and 80% mortality of SS homozygotes cause only about 10%, 70% and 50% mortality of F1 larvae, respectively. For the sake of simulations shown in figure 3,

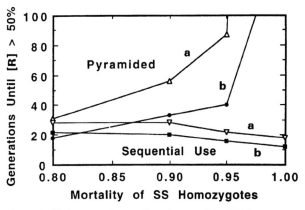

Figure 3. Influence of mortality of susceptible (SS) homozygotes on the evolution of resistance to transgenic crops for one toxin used alone (toxin A or B) or two toxins 'pyramided' in the same plant (A plus B), assuming that there is no cross-resistance. For simplicity, it is assumed that: (i) the mortalities of the single heterozygotes ($R_aS_aS_bS_b$ or $S_aS_aR_bS_b$) would be 70% when tested against just the one toxin to which they are resistant at SS mortalities of 90–100%, and 50% when SS mortality is 80%; (ii) mortalities of the homozygous susceptible genotypes are the same for both genes; and (iii) 20% (curve 'a') or 10% (curve 'b') of the eggs in each generation are laid on non-transgenic hosts. To more easily make comparisons across one- and two-toxin strategies, the mortalities for susceptible homozygotes are given in terms of what would be observed if the larvae were exposed to just one toxin at a time.

I assumed, for simplicity, that the mortality of RS heterozygotes was 70% for cases of susceptible mortality ranging from 90 to 100%, but to conservatively avoid making resistance excessively recessive, RS mortality was 50% when SS mortality dropped to 80%.

For single-toxin plants, the initial frequency of resistance alleles is less important to durability than the mortality of heterozygotes (Roush 1994, 1997a,b). However, pyramids are considerably more effective when resistance frequencies are low, provided that susceptible homozygotes are all killed by each of the toxins used separately (figure 4, compare curves labelled 'a' for sequential introductions and pyramids). However, even a lower initial frequency of the resistance allele does not significantly help in cases where transgenic plants cause only the poor mortality of susceptible homozygotes. For example, where the mortality of susceptible homozygotes and heterozygotes is only 80% and 50%, respectively (the point at 80% mortality on curve 'b' in figure 3), even with an initial allele frequency of 10^{-5}, the pyramiding strategy is slightly worse than a sequential release strategy, with resistance in 32 and 35 generations, respectively. As in figure 3, not much is lost by pyramiding; on the other hand, much can be gained from pyramiding if the mortality of susceptible insects is consistently greater than 95%, especially if the 'pyramided' varieties are released while initial resistance allele frequencies are still low (figure 4).

(e) *Effect of seed-line purity on pyramids*

As already seen, the pyramiding strategy relies on the high mortality of susceptible insects for each of the toxins used in the pyramids. However, because of practical

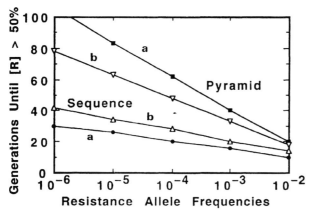

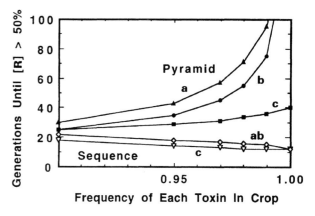

Figure 4. Effect of initial allele frequencies on the evolution of resistance for sequential and pyramided deployment of two toxins. A 10% refuge is assumed, with either 70% mortality of RS heterozygotes and 100% mortality of SS homozygotes (curve 'a'), or 70% for RS and 90% for SS for each toxin (curve 'b'). For comparative purposes, the curves labelled 'b' are an expansion across a wider initial frequency of the resistance allele of the points given for curve 'b' at 90% mortality of SS homozygotes (and an initial frequency of 10^{-3}) in figure 3.

Figure 5. Effect of seed-line impurity on durability of pyramids compared with the sequential use of the same toxins. Mortality of heterozygotes is 70% and 100% for susceptible homozygotes for each toxin for curves 'a' and 'b', but 5% of susceptible homozygotes survive for the curves labelled 'c'. For pyramids, either 100% of the crop carries toxin A, but the frequency of B is varied across a range of 90–100% (curve 'a'), or both toxins are allowed to vary in frequency from 90 to 100%, with their joint occurrence assumed to be random (curves 'b' and 'c'). For sequential introductions, both cultivars have the same levels of seed purity.

problems in incorporating two genes into the same cultivar, there may be some impurity of seed lines, such that some plants carry only one or the other of the toxin genes. For example, when the frequency of each toxin in the crop is 95% and segregating is random, the frequency of plants with both A and B is 0.9025, that of plants with A or B only is 0.095 (0.0475 each), and that of plants with neither A nor B is 0.0025. This would provide considerable opportunity for selection for resistance to either A or B without protection from the other toxin, and faster resistance (figure 5). Seed producers indicate that at least 97% purity is expected, which should ensure considerable benefits from pyramiding, but the increase in durability from improved purity would seem to justify the extra effort whenever there is 100% mortality of susceptible insects from each of the toxins deployed alone (figure 5, curves 'a' and 'b'). For example, when both toxins are present at a frequency of only 99%, resistance evolves in 75 generations, only half that when both toxins are represented in every plant (160 generations). On the other hand, if even 5% of the susceptible insects survive each toxin, there is still an advantage to pyramiding, but comparatively little benefit from increased seed purity (figure 5, 'c' curves).

(f) *Effect of linkage between resistance loci on pyramids*

Close chromosomal linkage between two resistance loci decreases the benefits of a mixture (Mani 1985). However, in the absence of other factors that limit the effectiveness of a pyramid, not even close linkages reduce the durability of pyramids to that of sequential introductions (figure 6).

(g) *Cross-resistance among Bt toxins*

Because the Bt Cry endotoxins are at least superficially similar to one another, it is possible that resistance genes will evolve that can overcome both Bt genes to be used in the pyramid, even if they have different binding sites. It is

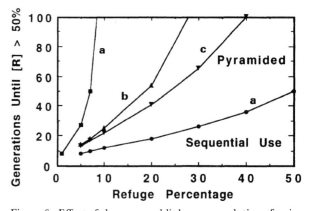

Figure 6. Effect of chromosomal linkage on evolution of resistance for pyramids and sequential introduction of the same toxins. Curve 'a' assumes no linkage and is the same as 'a' in figure 2; curve 'b' is same as 'a' but with recombination reduced to 10% from 50%. In 'c', recombination is only 1%.

widely assumed that such a resistance mechanism would be degradative and have a dominant expression, even though no such genes have yet been described (as discussed below). Strictly speaking, it is not dominance *per se* (the resemblance to resistant parents) that is important, but the extent to which heterozygotes survive. Thus, whether degradative or dominant, any genes that provide greater than 10% survival on the transgenic plants would cause resistance rather quickly (figure 1), and the pyramiding strategy will be much less effective than expected when compared with the simulation results given so far. However, the more important question is whether it would be better to sequentially deploy the genes given the possibility of cross-resistance, i.e. what is the relative effect of cross-resistance on both options? For example, a cross-resistance gene conferring 30% survival in heterozygotes with a high initial frequency would cause the rapid failure of a pyramid, but it would also do the same for a sequential deployment strategy (as will be subsequently illustrated).

Simulation modelling can be used to identify the characteristics of cross-resistance genes and the conditions of toxin expression under which sequential deployment would be more effective than pyramiding. To model cross-resistance, an additional locus was added to the model described to this point. One resistance locus continued to confer resistance to toxin A, and one to toxin B, with the survival of the pyramid being the product of survival values for each of the toxins when used alone. The third locus (C) confers resistance to both toxins, and is added to the survival conferred by the first two loci, up to a maximum of 100% survival. In considering the literature on pesticide resistance mechanisms in the broadest sense, it is hard to imagine any gene could confer cross-resistance without also conferring resistance to each Bt toxins when used alone.

If the cross-resistance gene, R_c, has fitnesses and initial gene frequencies greater than or equal to those of the specific genes, R_a and R_b, selection for resistance to any one toxin deployed alone occurs just about as quickly for pyramids. This is simply because the cross-resistance gene has the same advantages for resistance to the single toxins as do the specific resistance genes. For example, let us assume that all of the resistance genes have the same initial frequency (10^{-3}), a refuge of 10%, and the same fitnesses for all similar genotypes in the presence of the appropriate toxins. Specifically, assume that the susceptible S_aS_a and S_cS_c homozygotes all die when feeding on plants with toxin A, and all S_bS_b and S_cS_c die on plants with toxin B; but only 70% of R_aS_a and R_cS_c heterozygotes die when feeding on plants with toxin A and 70% of R_bS_b and R_cS_c on plants with toxin B. Simulations show that when the toxins are pyramided, the frequency of R_c reaches 64% after six generations of selection, but that the frequencies of R_a and R_b are essentially unchanged. In contrast, if only toxin A is deployed, frequencies of both R_a and R_c reach 50% after seven generations (figure 7, lower pair of curves, second pair of points from the left). If the R_cS_c heterozygotes suffer only 60% mortality, simulations predict that the frequency of R_c would exceed 50% in only five generations. However, the frequency of R_c would also reach 59% by six generations if either toxin A or B was deployed alone (figure 7, lowest left pair of points).

When the R_cS_c heterozygotes suffer 80% mortality and the R_aS_a and R_bS_b heterozygotes only 70%, the initial use of toxin A alone selects for the R_a allele ('R_a (seq)' in figure 8) a little more rapidly than the R_c allele ('R_c (seq)'), but R_c is still strongly selected when both resistance alleles are rare. The rate of increase of R_c slows after the frequency of R_a exceeds 10%, owing to the reduced average contribution of R_c to survival, but then increases again when selection is reintroduced by the switch to toxin B. However, because R_c is already so common from the selection with toxin A, resistance to toxin B by R_c appears by generation 8, even though the toxin B-specific resistance allele R_b is still uncommon. Thus, there is widespread resistance to both toxins by generation 8. If the two genes were pyramided ('pyr' in figure 8), cross-resistance does cause the pyramid to fail first (the increase shown is almost entirely to R_c), but only a generation earlier (figure 8).

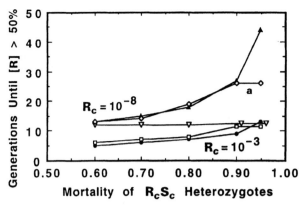

Figure 7. Influence of a gene for cross-resistance, R_c, on the durability of pyramided (filled symbols) and sequentially introduced toxins (open symbols). For a refuge of 10%, R_aS_a and R_bS_b mortality is 70%, and S_aS_a and S_bS_b is 100%, except in curve 'a', where R_aS_a and R_bS_b mortality is 95%. The initial frequency for R_a and R_b is 10^{-3} for toxins A and B, but the initial frequency for R_c is either 10^{-3} (lower pair of curves) or 10^{-8} (upper pair of curves and flat line of open triangles).

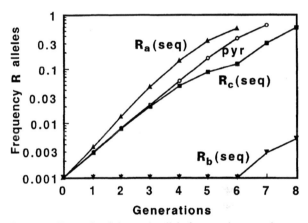

Figure 8. Example of change in allele frequencies over time for a gene conferring cross-resistance and two alleles causing more specific resistance; a more detailed look at the case when the mortality of R_cS_c is 80% and the initial frequency for all resistance alleles, R_a, R_b and R_c, is 10^{-3} (as for the third pair of points from the left in the bottom curve of figure 7). 'pyr' stands for pyramid, 'seq' for sequence.

These results should not be surprising. The cross-resistance gene has selective advantages no different than any other single gene without regard to whether there are one or two toxin genes present, and any gene conferring more than 10% survival in heterozygotes will cause resistance quickly in the absence of the benefits of pyramiding (figure 1). For sake of reference, in the absence of cross-resistance, sequential use under these assumptions would be expected to last for 12 generations and pyramids for 160 (results outlined in figures 2 and 3). In the presence of the cross-resistance gene, all-use strategies fail quickly; there is no advantage to pyramids, but little loss either. In the absence of cross-resistance, there is a ten-fold advantage to pyramiding.

These simulations imply that if broad cross-resistance genes exist, they must have frequencies and/or fitnesses lower than those of the Bt-resistance genes that have already been described. Otherwise, such broad cross-resistance genes should already have been encountered, but they have not. In at least four intensively studied cases of Bt resistance in the diamondback moth, resistance does not extend to Cry 1C (Tabashnik *et al.* 1997*b*, this issue; Tang *et al.* 1996), which would have been the best candidate to pyramid with Cry 1A. Although there were early reports of broad cross-resistance in *H. virescens* (Gould *et al.* 1992), subsequent studies have shown that the major locus for resistance to Cry 1A did not confer a significant effect on resistance to Cry 2A (Heckel *et al.* 1996).

To explore this further, it was assumed that the major cross-resistance locus had an initial frequency of 10^{-8}, but all else remained the same (figure 7, except for curve 'a'). Here again, there was little difference between pyramiding and sequential release or else pyramiding held the advantage (figure 7). When the initial frequency of R_c is 10^{-8}, the pyramiding curve is the same whether the mortalities of R_aS_a and R_bS_b are 70% or 95% (curve 'a'), because resistance for the pyramid is driven entirely by the survival of R_cS_c. When the initial frequency of R_c is 10^{-8} and mortality of R_cS_c is greater than 60%, R_c never reaches 50% frequency in sequential use, so the time for resistance is a flat 12 generations. Note the flattening of the 'a' sequences curve (just above the 'a' itself). In this case, the survival and initial frequency of R_cS_c are so low that it never significantly contributes to resistance in the sequence; resistance is simply due to R_a and R_b.

Because the characteristics, fitnesses and frequencies of cross-resistance genes cannot yet be fully anticipated, it may not be possible to completely discount their impacts on pyramided varieties. However, the conditions under which cross-resistance is most likely to cause failures of pyramids faster than sequential deployment, seem to be (i) when the mortalities of resistance heterozygotes for toxins A and B are sufficiently high enough (95% or higher) for there to be an appreciable delay of resistance even when the toxins are deployed alone (as in figure 1), and simultaneously (ii) where pyramids would not be strongly favoured even in the absence of cross-resistance, particularly when there is survival of susceptible homozygotes to the toxins when they are deployed alone (as in figure 3).

(h) *Differences with pesticide mixtures*

Contrary to the popular myth, there is no good experimental evidence that insecticide mixtures help to manage resistance (Tabashnik 1989). How is it that pyramiding of transgenic plants can succeed where pesticide mixtures have failed? Pesticide mixtures, in effect, too often occupy those regions of 'parameter space' where two-toxin strategies are no better than single-toxin strategies (as shown in the lower left-hand area of figure 3). As a result of incomplete coverage and residue decay, the mortality of susceptible homozygotes is rarely consistently high enough for pesticide mixtures to be effective. As an illustration from laboratory experiments, selection against the Indian meal moth with mixtures of toxins at concentrations that initially allowed 19% of the insects to survive, and rarely killed more than 75% of the selected line, produced resistance fairly rapidly, with little delay compared with the use of individual toxins (McGaughey

& Johnson 1992), just as would be predicted from these models. Achieving the high mortality of susceptible homozygotes is a key problem for pyramids, but at least the current cultivars meet or come close to this standard for the targeted pests. Another problem is that it seems unlikely that the mortality of heterozygotes is high in the field for most cases of pesticide resistance, where resistance is so often dominant in the field (Roush & Daly 1990). Yet another key difference is that whereas there are relatively low economic and environmental costs to pyramids, the use of pesticide mixtures requires higher pesticide application costs and increased risks for the environment, especially in terms of effects on natural enemies of pests (Tabashnik 1989).

6. CONCLUSIONS

The pyramiding of toxin genes offers what appears to be the most effective way to manage resistance to Bt and other insecticidal transgenic toxins. Obtaining consistently high mortality of susceptible homozygotes is a major limitation to the durability of pyramids, but it is a factor that can readily and easily be tested before any prospective release. A major limitation for single-toxin plants, low mortality of heterozygous insects, cannot be easily tested before release, because it requires anticipating all manner of resistance alleles that may occur at low frequency in the field. Cross-resistance has long been a concern for the use of pesticide mixtures and pyramids, but this paper suggests that the risks have been greatly overestimated. If cross-resistance occurs, pyramids seem unlikely to do much worse than sequential releases of the same toxin genes; on the other hand, in the absence of cross-resistance, pyramids may cause a great delay of resistance.

Pyramids have the potential to greatly reduce refuge requirements for successful resistance management from perhaps 30–40% down to perhaps 10% (figure 2). However, small refuges remain risky (as when mortalities of heterozygotes are lower than expected, e.g. case 'b' in figure 2). The more prudent way to deploy transgenic crops remains to keep refuges as large as is economically feasible. To prevent economic losses to these refuges, other non-insecticidal control techniques (e.g. pheromone disruption of mating, classically bred resistance, suppression of overwintering stages of insects through stalk or soil disruption, and crop rotation) should be used to manage population growth across the entire system of transgenic and non-transgenic plants (Roush 1997*a*).

The motto of the Royal Society is *'Nullius in verba'*, which can be translated as 'take nobody's word for it' and expressed the determination of early Fellows of the Society to verify all statements with an appeal to facts. In this case, I encourage interested readers to further investigate the points raised with additional modelling and experiments. It is widely thought that Bt transgenic crops are at risk from the rapid evolution of resistance. It is therefore important for the scientific community to rapidly come to a consensus about the best tactics for resistance management and to lobby for their implementation.

For the sake of brevity, I have often cited reviews rather than their original papers, but I thank T. J. Higgins, T. N. Hanzlik,

R. Wu, M. Peferoen & J. Van Rie, and G. P. Fitt for information on their work on alpha-amylase inhibitors in peas, *Helicoverpa* stunt virus, proteinase inhibitors in rice, Bt toxins on corn borers, and Bt expression in transgenic cotton, respectively. I also thank B. Tabashnik, F. Gould, D. Heckel, F. Perlak, G. Fitt and M. Caprio for discussions and correspondence on topics discussed here, and A. M. Shelton, J. D. Tang, P. Davis and G. P. Fitt for research collaborations.

REFERENCES

Andow, D. A. & Hutchison, W. D. 1998 Bt-corn resistance management. In *Now or never: serious new plans to save a natural pest control* (ed. M. Mellon & J. Rissler), pp. 19–66. Cambridge, MA: Union of Concerned Scientists.

Bennett, J., Cohen, M. B., Katiyar, S. K., Ghareyazie, B. & Khush, G. S. 1997 Enhancing insect resistance in rice through biotechnology. In *Advances in insect control: the role of transgenic plants* (ed. N. Carozzi & M. Koziel), pp 75–93. London: Taylor and Francis.

Carozzi, N. & Koziel, M. 1997 *Advances in insect control: the role of transgenic plants* London: Taylor and Francis.

Comins, H. 1986 Tactics for resistance management using multiple pesticides. *Agric. Ecosyst. Environ.* **16**, 129–148.

Curtis, C. F. 1985 Theoretical models of the use of insecticide mixtures for the management of resistance. *Bull. Entomol. Res.* **75**, 259–265.

Feldman, J. & Stone, T. 1997 The development of a comprehensive resistance management plan for potatoes expressing the Cry 3A endotoxin. In *Advances in insect control: the role of transgenic plants* (ed. N. Carozzi & M. Koziel), pp. 49–61. London: Taylor and Francis.

Fitt, G. P., Mares, C. L. & Llewellyn, D. J. 1994 Field evaluation and potential ecological impact of transgenic cottons (*Gossypium hirsutum*) in Australia. *Biocontrol Sci. Technol.* **4**, 535–548.

Georghiou, G. P. 1990 The effect of agrochemicals on vector populations. In *Pesticide resistance in arthropods* (ed. R. T. Roush & B. E. Tabashnik), pp. 183–202. New York: Chapman & Hall.

Gould, F. 1994 Potential and problems with high-dose strategies for pesticidal crops. *Biocontrol Sci. Technol.* **4**, 451–461.

Gould, F. 1986a Simulation models for predicting durability of insect-resistant germplasm: a deterministic diploid, two locus model. *Environ. Entomol.* **15**, 1–10.

Gould, F. 1986b Simulation models for predicting durability of insect-resistant germplasm: Hessian fly (Diptera: Cecidomyiidae)-resistant winter wheat. *Environ. Entomol.* **15**, 11–23.

Gould, F. 1998 Sustainability of transgenic insecticidal cultivars: integrating pest genetics and ecology. *A. Rev. Entomol.* **43**, 701–726.

Gould, F. &. Tabashnik, B. E. 1998 Bt-cotton resistance management. In *Now or never: serious new plans to save a natural pest control* (ed. M. Mellon & J. Rissler), pp. 67–105. Cambridge, MA: Union of Concerned Scientists.

Gould, F., Martinez-Ramirez, A., Anderson, A., Ferre, J., Silva, F. J. & Moar, W. F. 1992 Broad-spectrum resistance to *Bacillus thuringiensis* toxins in *Heliothis virescens*. *Proc. Natn. Acad. Sci. USA* **89**, 7986–7988.

Gould, F., Follet, P., Nault, B. & Kennedy, G. G. 1994 Resistance management strategies for transgenic potato plants In *Advances in potato pest biology and management* (ed. G. W. Zehnder, M. L. Powelson, R. K. Jansson & K. V. Raman), pp. 255–277. St Paul, MN: American Phytopathological Society Press.

Gould, F., Anderson, A., Jones, A., Sumerford, D., Heckel, D. G., Lopez, J., Micinski, S., Leonard, R. & Laster, M. 1997 Initial frequency of alleles for resistance to *Bacillus thuringiensis* toxins in field populations of *Heliothis virescens*. *Proc. Natn. Acad. Sci. USA* **94**, 3519–3523.

Heckel, D. G., Gahan, L. C., Gould, F. & Tabashnik, B. E. 1996 Mapping major and minor loci conferring resistance to Bt toxins in Lepidoptera. In *Proceedings of the second Pacific Rim conference on biotechnology of* Bacillus thuringiensis *and its impact to the environment*, Nov 4–8, pp. 468–480. Chiang Mai, Thailand: Entomology and Zoology Assocation.

McGaughey, W. H. & Johnson, D. E. 1992 Indian meal moth (Lepidoptera: Pyralidae) resistance to different strains and mixtures of *Bacillus thuringiensis*. *J. Econ. Entomol.* **85**, 1594–1600.

Mallet, J. & Porter, P. 1992 Preventing insect adaptation to insect-resistant crops: are seed mixtures or refugia the best strategy? *Proc. R. Soc. Lond.* B **250**, 165–169.

Mani, G. S. 1985 Evolution of resistance in the presence of two insecticides. *Genetics* **109**, 761–783.

Maxwell, F. G. & Jennings, P.R. (eds) 1980 *Breeding plants resistant to insects*. New York: Wiley.

Metz, T. D., Roush, R. T., Tang, J. D., Shelton, A. M. & Earle, E. D. 1995 Transgenic broccoli expressing a *Bacillus thuringiensis* insecticidal crystal protein: implications for pest resistance management strategies. *Mol. Breeding* **1**, 309–317.

Painter, R. H. 1951 *Insect resistance in crop plants*. New York: Macmillan.

Perez, C. J. & Shelton, A. M. 1997 Resistance of *Plutella xylostella* (Lepidoptera: Plutellidae) to *Bacillus thuringiensis* Berliner in Central America. *J. Econ. Entomol.* **90**, 87–93.

Purcell, J. P. 1997 Cholesterol oxidase for the control of boll weevil. In *Advances in insect control: the role of transgenic plants* (ed. N. Carozzi & M. Koziel), pp. 95–108. London: Taylor and Francis.

Roush, R. T. 1989 Designing resistance management programs: how can you choose? *Pestic. Sci.* **26**, 423–441.

Roush, R. T. 1994 Managing pests and their resistance to *Bacillus thuringiensis*: can transgenic crops be better than sprays? *Biocontrol Sci. Technol.* **4**, 501–516.

Roush, R. T. 1996 Can we slow adaptation by pests to insect transgenic crops? In *Biotechnology and integrated pest management* (ed. G. Persley), pp. 242–263. London: CABI.

Roush, R. T. 1997a Managing resistance to transgenic crops. In *Advances in insect control: the role of transgenic plants* (ed. N. Carozzi & M. Koziel), pp. 271–294. London: Taylor and Francis.

Roush, R. T. 1997b Bt-transgenic crops: just another pretty insecticide or a chance for a new start in resistance management? *Pestic. Sci.* **51**, 328–334.

Roush, R. T. & Daly, J. C. 1990 The role of population genetics in resistance research and management. In *Pesticide resistance in arthropods* (ed. R. T. Roush & B. E. Tabashnik), pp. 97–152. New York: Chapman & Hall.

Roush, R. T. & Shelton, A. M. 1997 Assessing the odds; the emergence of resistance to Bt transgenic plants. *Nature Biotech.* **15**, 816–817.

Schroeder, H. E., Gollasch, S., Moore, A., Tabe, L. M., Craig, S., Hardie, D. C., Chrispeels, M. J., Spencer, D. & Higgins, T. J. V. 1995 Bean alpha-amylase inhibitor confers resistance to the pea weevil (*Bruchus pisorum*) in transgenic peas (*Pisum sativum* L.). *Plant Physiol.* **107**, 1233–1239.

Schuler, T. H., Poppy, G. M., Kerry, B. R. & Denholm, I. 1998 Insect-resistant transgenic plants. *Trends Biotechnol.* **16**, 169–175.

Shelton, A. M., Tang, J. D., Earle, E. D. & Roush, R. T. 1998 Can we manage resistance to Bt-engineered plants? Results of greenhouse and field tests. In *Proceedings of the Sixth Australian Applied Entomological Research Conference, Brisbane, Australia, October 1998.* (In the press.)

Tabashnik, B. E. 1989 Managing resistance with multiple pesticide tactics: theory, evidence, and recommendations. *J. Econ. Entomol.* **82**, 1263–1269.

Tabashnik, B. E. 1994*a* Evolution of resistance to *Bacillus thuringiensis*. *A. Rev. Entomol.* **39**, 47–79.

Tabashnik, B. E. 1994*b* Delaying insect adaptation to transgenic crops: seed mixtures and refugia reconsidered. *Proc. R. Soc. Lond.* B **255**, 7–12.

Tabashnik, B. E., Schwartz, J. M., Finson, N. & Johnson, M. W. 1992 Inheritance of resistance to *Bacillus thuringiensis* in diamondback moth (Lepidoptera: Plutellidae). *J. Econ. Entomol.* **85**, 1046–1055.

Tabashnik, B. E., Liu, Y.-B., Finson, N., Masson, L. & Heckel, D. G. 1997*a* One gene in diamondback moth confers resistance to four *Bacillus thuringiensis* toxins. *Proc. Natn. Acad. Sci. USA* **94**, 1640–1644.

Tabashnik, B. E., Liu, Y.-B, Malvar, T., Heckel, D. G., Masson, L., Ballester, V., Granero, F., Ménsua, J. L. & Ferré, J. 1997*b* Global variation in the genetic and biochemical basis of diamondback moth resistance to *Bacillus thuringiensis*. *Proc. Natn. Acad. Sci. USA* **94**, 12 780–12 785.

Tang, J. D., Shelton, A. M., Van Rie, J., De Roeck, S., Moar, W. J., Roush, R. T. & Peferoen, M. 1996 Toxicity of *Bacillus thuringiensis* spore and crystal protein to the resistant diamondback moth (*Plutella xylsotella*). *Appl. Environ. Microbiol.* **62**, 564–569.

Tang, J. D., Gilboa, S., Roush, R. T. & Shelton, A. M. 1997 Inheritance, stability, and fitness of resistance to *Bacillus thuringiensis* in a field colony of *Plutella xylostella* (L.) (Lepidoptera: Plutellidae) from Florida. *J. Econ. Entomol.* **90**, 732–741.

Myths, models and mitigation of resistance to pesticides

Marjorie A. Hoy

Department of Entomology and Nematology, PO Box 110620, 970 Hull Road, University of Florida, Gainesville, FL 32611-0620, USA
(mahoy@gnv.ifas.ufl.edu)

Resistance to pesticides in arthropod pests is a significant economic, ecological and public health problem. Although extensive research has been conducted on diverse aspects of pesticide resistance and we have learned a great deal during the past 50 years, to some degree the discussion about 'resistance management' has been based on 'myths'. One myth involves the belief that we can manage resistance. I will maintain that we can only attempt to mitigate resistance because resistance is a natural evolutionary response to environmental stresses. As such, resistance will remain an ongoing dilemma in pest management and we can only delay the onset of resistance to pesticides.

'Resistance management' models and tactics have been much discussed but have been tested and deployed in practical pest management programmes with only limited success. Yet the myth persists that better models will provide a 'solution' to the problem. The reality is that success in using mitigation models is limited because these models are applied to inappropriate situations in which the critical genetic, ecological, biological or logistic assumptions cannot be met. It is difficult to predict in advance which model is appropriate to a particular situation; if the model assumptions cannot be met, applying the model sometimes can increase the rate of resistance development rather than slow it down.

Are there any solutions? I believe we already have one. Unfortunately, it is not a simple or easy one to deploy. It involves employing effective agronomic practices to develop and maintain a healthy crop, monitoring pest densities, evaluating economic injury levels so that pesticides are applied only when necessary, deploying and conserving biological control agents, using host-plant resistance, cultural controls of the pest, biorational pest controls, and genetic control methods. As a part of a truly multi-tactic strategy, it is crucial to evaluate the effect of pesticides on natural enemies in order to preserve them in the cropping system. Sometimes, pesticide-resistant natural enemies are effective components of this resistance mitigation programme. Another name for this resistance mitigation model is integrated pest management (IPM). This complex model was outlined in some detail nearly 40 years ago by V. M. Stern and colleagues.

To deploy the IPM resistance mitigation model, we must admit that pest management and resistance mitigation programmes are not sustainable if based on a single-tactic strategy. Delaying resistance, whether to traditional pesticides or to transgenic plants containing toxin genes from *Bacillus thuringiensis*, will require that we develop multi-tactic pest management programmes that incorporate all appropriate pest management approaches. Because pesticides are limited resources, and their loss can result in significant social and economic costs, they should be reserved for situations where they are truly needed—as tools to subdue an unexpected pest population outbreak. Effective multi-tactic IPM programmes delay resistance (=mitigation) because the number and rates of pesticide applications will be reduced.

Keywords: evolution; resistance models; resistance management; pesticide resistance; integrated pest management; pesticide selectivity

1. INTRODUCTION

Resistance to pesticides in arthropod pests is a significant economic, ecological and public health problem (Georghiou & Saito 1983; Georghiou 1986; National Academy of Sciences 1986; Roush & Tabashnik 1990; Denholm *et al.* 1992; McKenzie 1996). More than 500 arthropod species have become resistant to insecticides and acaricides, with many species having become resistant to the major classes of such products.

In this essay I will address three myths that must be dispelled before we can adopt a more effective paradigm for reducing the effects of pesticide resistance. One myth involves terminology that affects how we think about the problem, another involves the effectiveness and appropriateness of 'management' models, and a third myth involves our reluctance to recognize that resistance to pesticides and other xenobiotics is an evolutionary response to stress that will remain a persistent problem.

2. TERMINOLOGY MYTHS

'Managing' resistance is, in my opinion, an inappropriate term for what we can achieve and distorts our perception of our true objective. According to Webster's dictionary (2nd college edition, 1982), 'managing' is defined as 'to control the movement or behavior of; handle; manipulate; to have charge of'. I think that 'having charge of, or controlling' resistance is a myth. At best we can delay the onset of resistance. A better term might be 'mitigate', which is defined as 'to make or become milder, less severe, less rigorous or less painful; to moderate'. I will argue that our more realistic goal is to mitigate resistance.

3. MODELS AND MYTHS

Scientists have attempted to model pesticide resistance in a variety of ways and the models have been extremely helpful in clarifying issues and concepts (Tabashnik 1990). The models can be classified by the basic assumptions employed, the modelling approach taken, the variables considered, and the problem addressed. Analytical, simulation, optimization and empirical models have been developed, but each has limitations and/or assumptions that are not always recognized. These limitations and assumptions severely restrict the generality and applicability of the models to resolving real-world problems.

(a) Analytical models

Analytical models, which attempt to analyse general trends using a simple mathematical description without providing realistic details, are relatively simple and '... seek to define fundamental principles' (Tabashnik 1990). Analytical models usually '... assume simple population dynamics with discrete generations and no age structure. Population growth is usually determined by some form of the logistic equation' (Tabashnik 1990).

How realistic are these assumptions and what effect would violation of these assumptions have on the outcome of the model? We know that relatively few arthropods have discrete generations; most of those that are prone to developing resistance (spider mites, aphids and whiteflies) are multivoltine, and have overlapping generations.

(b) Simulation models

Simulation models are more complex and realistic than analytical models because they attempt to incorporate details of the biology, behaviour and ecology of the population and often contain complex population dynamics, including age structure, overlapping generations and temporal and spatial variation in pesticide dose (Tabashnik 1990). Simulation models can be used to evaluate different options for mitigating resistance by including empirical data in the parameters included in the model. These parameters also can be varied in a systematic way to determine how important each is. Yet simulation models have serious limitations, as well.

Collaborations with Michael Caprio and Bruce Tabashnik allowed me to recognize additional issues important to resistance development (Caprio & Hoy 1994, 1995; Caprio et al. 1991). Our goal was to increase frequencies of pesticide resistance alleles in natural enemy populations. The development of simulation models was useful and intellectually stimulating, but it became very clear that the details of the population biology and ecology of

species sometimes had a pronounced influence on the rate of resistance development. For example, premating isolation and metapopulation structure can influence rate of resistance development in unexpected ways.

Premating isolation in diplodiploid and in haplodiploid species affected the rate of resistance development in sometimes counterintuitive ways (Caprio & Hoy 1995). The amount of mating bias (preference of females to mate with males of the same genotype) determined the rate of establishment when resistant individuals comprised 10% of the population. Interactions between mating bias, degree of dominance and diploidy state also were significant.

Population structure also may influence the rate of resistance development (Caprio & Hoy 1994). A stochastic metapopulation model investigating the establishment of a pesticide-resistant strain of predatory mite found that metapopulation dynamics increased local homozygosity within predator patches, and thus accelerated resistance development most when the resistance mechanism was recessive. Metapopulation dynamics also were important in inducing genetic bottlenecks by high rates of overwintering mortality, which synchronized loss of rare alleles in small populations. We concluded that the mitigation tactics that '... reduce the pest species' population size at critical periods such as overwintering may limit the potential of those populations to maintain resistance alleles' (Caprio & Hoy 1994).

What happens if we make simulation models more complex and more like the real world? Can we really capture all the essential biological, ecological and behavioural aspects of a particular pest species? Do we know enough about these details for many species? Is each species or population unique? Jaffee et al. (1997) pointed out that 'One of the most important criticisms to the use of models in biology, and in explaining genetic resistance in particular, is that biological and ecological systems are rather complex, and that simple models ignore that many relevant biological phenomena are emergent properties from complex interactions.' The concept of emergent properties suggests that we are unlikely to know enough to adequately model real species in sufficient detail. Jaffee et al. (1997) concluded 'This criticism is difficult to refute as evidence of the emergence of unexpected properties from complex system simulations is mounting.'

Despite the concerns of Jaffee et al. (1997) about complexity and emergent properties, they developed a complex model incorporating the effect of various selection pressures on 17 different genes evolving simultaneously in a population. Their results confirmed previous findings that the likelihood of emergence of genetic resistance in a given population is related to many factors, including the size of the initial population, length of the treatment with pesticides, mutation rate, sexual strategy (sexual or asexual), application methods (rotations versus mixtures), timing of the pesticide applications, and residue length. They concluded that '... evolution under a complex assemblage of selection pressures is different from evolution driven by a single environmental factor such as a pesticide.' They further concluded that 'Emergence of genetic resistance is an irreversible process.' This implies that reversion is unlikely to provide

Table 1. *Important assumptions of resistance mitigation tactics investigated by models*

(Based on a review by Tabashnik (1990).)

model type	assumptions
mixtures	resistance to each product is monogenic no cross-resistance occurs between products in mixture resistant individuals are rare in the population products have equal persistence some of the population remain untreated (refuge) resistance is functionally recessive (only homozygotes survive exposure)
mosaics	susceptible individuals are maintained and able to move into surrounding patches may require negative cross-resistance or fitness costs associated with resistance
rotations	the frequency of individuals resistant to one product will decline during application of the alternative product, which is true if there is negative cross-resistance (rare), a substantial fitness cost associated with resistance, or immigration of susceptible individuals occurs
natural-enemy/pest system	food limitations are sufficient to constrain the ability of natural enemies to develop resistance in the field
high-dose strategy	assumes complete coverage, effective kill of all individuals, ignores negative effects on secondary pests, natural enemies, or the environment

an opportunity to reuse a product once resistance has developed.

(c) *Optimization models*

Optimization models evaluate, using dynamic programming techniques, which management strategy will maximize profit when pest susceptibility to a pesticide is considered a non-renewable natural resource. The goal is to balance the future cost of reduction in pest susceptibility with the present losses in crop yield due to the effects of the target pest. The details of the biology of the target species are usually simplified and considered a constraint to the model, which focuses on an economic analysis (Tabashnik 1990).

Because details of biology, ecology and behaviour of the pest are critically important to the development of resistance, as are the characteristics of the specific pesticide product, the rate at which resistance develops in specific situations may vary widely. Thus, optimization models may over- or underestimate the longevity of any product and lead to inaccurate predictions of the costs of losing a specific product.

(d) *Empirical models*

Empirical models are based on actual relationships among variables, with no assumptions made about the causal mechanisms. The models are derived from data and probably are appropriate only to the specific conditions of the observed populations (Tabashnik 1990). Thus, empirical models are least useful for developing a strategy for delaying resistance in an unknown situation if we assume that the important variables can vary between populations (mode of inheritance, cross-resistances, fitness costs, allele frequency and selection intensity).

(e) *Mitigation myths*

In my opinion, resistance mitigation investigated by models remains limited both by the objectives considered and the basic assumptions made (table 1). Even if we limit the discussion to the evolution of resistance by one pest, and the models include situations in which pesticides are applied in mixtures, rotations or mosaics, we remain uncertain whether to recommend alternation of different pesticides or to recommend mixtures as the best method for slowing the development of resistance.

(i) *Mixtures*

Mixtures of products are applied so that individuals are exposed simultaneously to more than one toxicant. Most models involving mixtures require a remarkable array of assumptions: that resistance to each product is monogenic, no cross-resistance occurs between the products used in the mixture, that resistant individuals are rare, the products have equal persistence, and that some of the population remains untreated (Tabashnik 1990). Another assumption is that resistance for each insecticide is functionally recessive so that only homozygous individuals survive. If resistance is not completely recessive, the rate of resistance development is increased. If the products are applied in the field in such a way as to vary the dosage each arthropod experiences, then some heterozygous individuals experiencing lower doses could survive, again speeding the rate of resistance development (Tabashnik 1990).

Few field experiments have been conducted with mixtures. How often can all these assumptions be met and what is the penalty if one or more is violated? I think the situations in which we know that cross-resistance will not occur are rare; we certainly would not expect there to be any degree of cross-resistance between products in different pesticide classes. However, cross-resistance between abamectin and pyrethroids has been reported (Lasota & Dybas 1991), and other examples of cross-resistances between different pesticide classes could be cited. Furthermore, how often is it appropriate to assume resistance is monogenic? When the genetics of resistance can be analysed carefully, as with *Drosophila melanogaster*, resistance often is found to be determined by genes located on more than one chromosome. Although a 'major' gene may determine the bulk of the resistance, several other loci also may contribute to the resistance. As geneticists, we should expect that resistance can develop in a variety of ways (either at different points in the biochemical pathway or through totally different mechanisms) (Scott 1990, 1995; Soderlund & Blomquist 1990). We know that different mutations (alleles) will vary in their effect, that different alleles will vary in their mode of inheritance (in a continuum from fully dominant to fully recessive), and that different populations may contain different resistance alleles or loci. 'Modifier genes' may affect the degree of resistance and the fitness of the organism.

(ii) *Rotations*

The hypothesis is that if two or more pesticides are alternated in time, each individual is exposed to only one material but the population experiences more than one product over time. The assumptions include: the frequency of individuals that are resistant to one product will decline during the application of the other product (which can occur if there is negative cross-resistance), a fitness cost associated with the resistance, or movement of susceptible individuals into the population.

The assumption that reduced fitness could be used in resistance mitigation programmes continues to be controversial and may have limited application. Resistance alleles do not always produce detectable levels of lowered fitness over a long period of time (e.g. Hoy & Conley 1989; Hoy 1990; Roush & Daly 1990). It is likely that natural selection will increase the number of 'modifying genes' that restore fitness to individuals carrying resistance alleles.

(iii) *Mosaics*

Mosaics are a spatial patchwork of pesticide applications so that different sites are treated with different pesticide products. Mosaic mitigation models require that susceptible individuals migrate into the treated area, or that negative cross-resistance occurs (which is rare), and that fitness costs are high. The size of the patches required will vary with the biology and ecology of the pest arthropod. Unfortunately, details of dispersal rate and distance vary by species, but often are unknown even for key pests.

(f) *Experimental validation of models*

Relatively few resistance 'mitigation' experiments have been conducted under realistic field conditions. Such experiments are expensive to conduct and require long time periods. Yet, without validation of mitigation models, we are left with little justification for recommending specific actions.

One experiment, in which resistance in the two-spotted spider mite, *Tetranychus urticae* Koch, was measured during seven years in southern Oregon pear orchards (Flexner *et al.* 1995), is particularly interesting because it illustrates some of the problems associated with the theoretical models and their limited application to specific field situations and pest populations. During the experiment,

five treatments were applied in replicated field plots twice a season: consecutive organotin use, consecutive hexythiazox use, alternation of both within year, between-year rotations of both organotins and hexythiazox, and a combination at half rates of both types of compound.

Flexner *et al.* (1995) concluded that 'Overall, use in the field was not extended by rotations or half-rate combinations compared with consecutive uses, but benefits from these programs may occur because of slow registration of new acaricides.' Thus, the rotations did not allow increased numbers of applications of a specific product to be made but, because the applications were made in different years, the products lasted longer, which could be useful if this time interval allowed for registration of new products. This might not provide a benefit if the pest population was already resistant to the new product, for they went on to state that 'Resistance to organotins conferred cross-resistance to hexythiazox.' Again, there was no reason, *a priori*, to assume cross-resistances between these two very different products. Flexner *et al.* (1995) also noted that 'cautions are needed before extending [our results] to other situations.' They were concerned that their relatively small plot sizes and the relatively high rate of immigration of susceptible individuals into the plots could have led them to overestimate the potential for resistance management. They were also concerned that although the population of *T. urticae* they worked with readily reverted to susceptibility when left unselected, such reversion does not occur in all populations of this mite. Third, they noted that the parameters used to define resistance (field failure and elevated LC_{50} values) are quite specific to the crop and cultivar.

Some of our most fundamental assumptions about resistance are being questioned. The assumption that resistance is preadaptive may be wrong in some cases. Devonshire & Field (1991) reviewed gene amplification and insecticide resistance in aphids and mosquitoes and speculated that insecticides might act to increase mutation rates, especially with regard to amplified resistance genes, although there are no data to support this at present. Another controversial issue is the possibility that resistance alleles are extremely rare and that resistant individuals may migrate much greater distances than expected, leading to the spread of resistance alleles around the world in a surprisingly short time (Guillemaud *et al.* 1996; Pasteur & Raymond 1996).

Tabashnik (1990) concluded that '... theoretical models and available data suggest that the effectiveness of mixtures, rotations, and mosaics requires special conditions that are not generally met in the field.' He further concluded that '... reducing pesticide use through integrated pest management may be more productive than attempts to optimize pesticide combinations.'

The resistance mitigation models developed to date generally rely on a single tactic (rotation of products or mixture of products or providing a patchwork of treated and untreated sites so that susceptible individuals persist). We have already learned that single-tactic mitigation models are unlikely to be sustainable over long periods of time. The proposed methods of mitigating resistance to *Bacillus thuringiensis* (Bt) toxins in transgenic crops (seed mixtures or pyramiding or refugia for susceptible individuals) are fundamentally single-tactic approaches (Roush 1996) and also are unlikely to be sustainable.

What do we know about mitigating resistance? I think we know a lot about what does not work. An evaluation of pest management programmes since the 1940s indicates that when pesticides are applied in a manner designed to achieve the elimination of a target pest, serious environmental and other problems usually ensue (National Academy of Sciences 1986; National Research Council 1989; Office of Technology Assessment 1992; Pimentel & Lehman 1993). The intensive and extensive use of pesticides to increase food production and improve human and animal health has failed to be sustainable. It seems likely that relying on transgenic crops that express high levels of a single toxin (a high-dose strategy) also will be a doomed strategy.

Although we can learn from studying pesticide resistance in ubiquitous pests in other geographic regions and thus be alerted to a potential problem, this is an inefficient and often inappropriate method for mitigating resistance in arthropods, especially if different species or different geographic populations develop resistance by different mechanisms. Geneticists know that resistance mechanisms may vary, their mode of inheritance may vary, and the degree of reduction in fitness associated with different alleles or loci may vary. Monitoring programmes are unlikely to be cost effective because it is difficult to sample rare individuals in natural populations (Brent 1986) and therefore they are best employed to document a problem once it has developed (Hoy 1992, 1995).

What is wrong with past resistance mitigation research? In my opinion, the problem is that resistance mitigation research and IPM research programmes usually have been considered different topics (Hoy 1992, 1995). As a result, an effective paradigm for resistance mitigation has not been adopted.

We have tried the simple approaches, in models and in experimental and operational programmes. The simple solutions and models fail owing to our lack of data on the true fitness costs, true selection intensity, mode of inheritance, dispersal rates and distance, and cross-resistance patterns of the resistance gene(s) in the target pest in the specific environment. Often, if one or more essential assumption is violated, the models (and the programmes) fail. Because it is nearly impossible to anticipate all key factors *a priori*, and it is difficult to obtain data in sufficient time, we need to adopt a different approach.

4. MULTI-TACTIC APPROACHES TO MITIGATING RESISTANCE

Multi-tactic approaches to mitigation of resistance are more robust and sustainable than single-tactic approaches. One multi-tactic resistance mitigation model was developed in 1959 and it continues to provide a sustainable solution to resistance if the principles developed then are adopted today. Stern *et al.* (1959) recognized that 'All organisms are subjected to the physical and biotic pressures of the environments in which they live, and these factors, together with the

genetic make-up of the species, determine their abundance and existence in any given area.' Stern *et al.* (1959) conducted their research to mitigate the problem of pesticide resistance in the spotted alfalfa aphid in California. They noted that 'Without question, the rapid and widespread adoption of organic insecticides brought incalculable benefits to mankind, but it has now become apparent that this was not an unmixed blessing.' The problems of resistance, secondary outbreaks of arthropods, resurgence of pest arthropods, toxic residues on food and forage crops, and hazards to insecticide handlers and persons, livestock and wildlife from contamination by pesticide drift '... have arisen from our limited knowledge of biological science; others are the result of a narrow approach to insect control' (Stern *et al.* 1959).

The integrated pest management programme developed by Stern *et al.* (1959) for alfalfa in California included a variety of tactics, including monitoring, assessing economic injury levels, using selective pesticide products, and integrating chemical and biological control. 'Chemical control of an arthropod pest is employed to reduce populations of pest species which rise to dangerous levels when the environmental pressures are inadequate. When chemicals are used, the damage from the pest species must be sufficiently great to cover not only the cost of the insecticidal treatment but also the possible deleterious effects ...' (Stern *et al.* 1959). They went on to state that 'Chemical control should be used only when the economic threshold is reached and when the natural mortality factors present in the environment are not capable of preventing the pest population from reaching the economic-injury level.' Stern & van den Bosch (1959) recognized that there was '... an imperative need for an insecticide that would give adequate aphid control and also allow the native predators to survive treatment' and concluded that 'The desirability of attaining a pest-control program in which chemical and biological control are as well integrated as possible is indisputable.'

A multi-tactic resistance mitigation model developed by Barclay (1996) compared the effects of combining methods of insect pest control on the rate of selection for resistance. He found that when two control methods are used in combination, selection for resistance against the two is a linear function if the two do not interact. If the two interact, the function may be sublinear or supralinear. He concluded that the '... control methods that appear least likely to encounter resistance are natural enemies and the use of pheromone traps for male annihilation. These should be integrated into a control program where possible to minimize the development of resistance to other control methods being used.'

Tabashnik & Croft (1985) demonstrated that evolution of resistance in pests was slowed when pesticide applications could be reduced because predators were maintained in the system (Tabashnik & Croft 1985; Tabashnik 1990). There is general agreement that reduced pesticide use is an essential element of any resistance mitigation programme (Croft 1990b; Tabashnik 1990; Metcalf 1994). Thus, the compatibility of pesticides and biological control agents is a crucial issue in pesticide resistance mitigation as well as effective IPM programmes.

Pesticide-resistant natural enemies are a special category of pesticide selectivity that can help to delay resistance in a multi-tactic IPM programme. Relatively few natural enemies have developed resistance to pesticides through natural selection, but several have been deployed in IPM programmes (Croft 1990a; Hoy 1990). Artificial selection of phytoseiid predators for pesticide resistance can be a practical and cost-effective tactic for the biological control of spider mites (Hoy 1990). However, development of pesticide-resistant natural enemies should not be considered before exploring other, less expensive options for IPM and pesticide resistance mitigation.

Multi-tactic resistance programmes have been suggested for managing resistance to crops containing Bt genes that include consideration of natural enemies. Hokkanen & Wearing (1995) suggested five tactics for Bt resistance mitigation in oilseed *Brassica*: (i) provide refugia for susceptible individuals; (ii) do not use pesticides that kill susceptible individuals of pests that are targets for control by the Bt gene; (iii) do not use pesticides that kill natural enemies of any of the pests in the crop; (iv) enhance natural control of pests by crop rotation or tillage practices; and (v) rotate between susceptible and resistant crop genotypes synchronously over large areas, while observing points (i–iv). Wearing & Hokkanen (1994) evaluated the potential for development of resistance to Bt genes inserted into apples and kiwi fruit in New Zealand. They suggested that '... the ecological characteristics of the pest provide strong natural mechanisms for retention of susceptibility.' The natural host range of the target species, the natural availability of refugia and the mobility and likely gene flow in the populations should promote susceptibility. 'Even in these circumstances, it is essential that Bt-apple and Bt-kiwi fruit in New Zealand are released into carefully managed IPM programmes, particularly avoiding pesticides toxic to [insects in] refugia, immigrants and natural enemies, and including mating disruption where required.'

5. LEGISLATIVE ISSUES

Nearly everyone will agree that reducing pesticide use is an effective resistance mitigation tactic (Croft 1990a; Tabashnik 1990). What has not been widely acknowledged is that resistance mitigation programmes also should include altering the way pesticides are developed and registered. Decisions on application rates and the numbers of applications per growing season should be made with the understanding that they affect the speed with which resistance will develop. In some cases, new products should not be registered for a specific crop because they are toxic to natural enemies and thus could disrupt effective IPM programmes already in place, which will speed the development of resistance in specific pests.

Adoption of new legislation requires that we admit that nearly all major insect and mite pests can develop resistances to all classes of pesticides given sufficient selection pressure over sufficient time. This important assumption may have exceptions, but the generalization is reasonable given the documented record of resistance development in arthropod pests during the past 50 years. Resistance to stress is a fundamental evolutionary response by living organisms and has been achieved by

diverse molecular methods (Scott 1995). On an evolutionary time-scale, we should expect insects to have evolved mechanisms to survive extreme temperatures, allelochemicals and other environmental stresses. Although new pesticide classes have been proclaimed to be a potential 'silver bullet', and not subject to resistance development, these hopes have been misplaced to date. It seems appropriate to assume that the development of resistance is nearly inevitable and the issue is not whether resistance will develop, but when.

There are increasing social, economic and ecological pressures to reduce pesticide use through legislative measures in the USA and to increase the use of non-chemical control tactics such as host-plant resistance, biorational methods, cultural controls and biological controls (National Research Council 1989; Office of Technology Assessment 1992; Lewis *et al.* 1997). There is an increasing interest on the part of research scientists, regulatory agencies, legislators and the public in using pesticides that are non-toxic to biological control agents and that have minimal impacts on the environment and human health. The issue of compatibility of pesticides with natural enemies and other non-chemical tactics is critical for improving pest management and environmental quality, and for mitigating resistance to pesticides in pest arthropods. Enhancing the compatibility of pesticides and biological control agents is complex and sometimes difficult (Croft 1990a; Hoy 1985, 1990; Hull & Beers 1985), but can pay handsome dividends in improved pest control (Metcalf 1994) and pesticide resistance mitigation (Tabashnik & Croft 1985).

If the pesticide registration process in the USA is changed, we can delay resistance as well as achieve improved IPM programmes (Hoy 1992, 1995). For example, some pesticides are relatively non-toxic to important natural enemies in cropping systems at low rates, but the recommended application rates are too high (Hoy 1985). Use at the high rates disrupts effective biological control, leading to additional pesticide applications, which exerts unnecessary selection for resistance in the pest. Under these circumstances, it may be appropriate for the label to contain two different directions for use; one rate could be recommended for the traditional strategy of relying solely on pesticides to provide control (although this is becoming a less viable option). A lower rate could be recommended for use in an IPM programme that employs effective natural enemies. This dual approach to labelling could reduce selection for resistance in both target and non-target pests in the cropping system.

Another innovation in pesticide registration in the USA would require that the toxicity of the pesticide be determined for a selected list of biological control agents in each cropping system. This information should be provided, either on the label or in readily available computerized databases, perhaps via the Internet. Without such information, pesticides are used that disrupt effective biological control agents, which often results in unnecessary use of pesticides. Enhancing biological control not only leads to improved pest management, but also is an essential tool in mitigating pesticide resistance.

How could information about the toxicity of pesticides to biological control agents best be made available? How

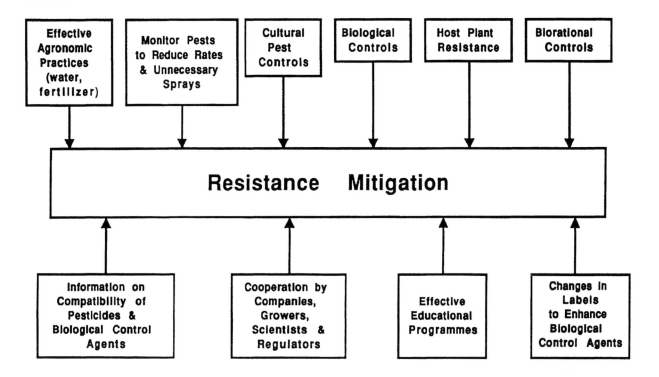

Figure 1. Effective resistance mitigation and integrated pest management: one and the same?

should bioassays be conducted to evaluate pesticide selectivity? There are no simple answers. Theiling & Croft (1988) and Croft (1990b) compiled an extensive set of data on the impact of pesticides on natural enemies, but additional data also are buried in publications or reports that are difficult to find. Unfortunately, even if the data can be found, it is not always easy to interpret bioassay data obtained by different scientists using different assay methods. Different bioassays can produce different conclusions about the toxicity of pesticides to natural enemies, and it is often difficult to predict the impact of pesticides under field conditions based on laboratory assays (Hoy 1990; Hassan *et al.* 1991; Robertson & Preisler 1992). Thus, the recommendation that labels or databases be developed with information on the impact of pesticides on natural enemies requires considerable discussion and additional research. Should pesticide companies conduct the research using standard bioassay methods? Should a consortium of pest management scientists conduct the assays? Who should pay for the research? What species of natural enemies should be tested? However, the concept is not new, and in Europe standardized bioassays already are being conducted on selected natural enemy species (e.g. Hassan *et al.* 1991; Oomen *et al.* 1994). Increased international consultation and cooperation between scientists, regulatory agencies and pesticide companies could resolve many of the questions raised above.

6. CONCLUSIONS

The mitigation of resistance in pest arthropods is a difficult and complicated business and is unlikely to be resolved by simple solutions. Mitigation of resistance to Bt toxins in plants also is unlikely to be easy or simple. It is doubtful that resistance to Bt-toxin genes can be prevented by stacking or pyramiding them in transgenic plants; in at least some arthropods a single gene confers resistances to multiple Bt toxins (Tabashnik *et al.* 1997). The deployment of crop plants with toxin genes that exert continuous selection pressure on both target and non-target arthropod populations, whether or not the target pest exceeds an economic injury level, is an unusually effective selection method. If the toxin genes are expressed at a sufficiently high level that arthropod populations are eliminated at least temporarily, both host-specific and generalist natural enemies will be unable to sustain themselves in the cropping system without their food, and this is a familiar scenario for inducing secondary pest outbreaks. Effective IPM programmes require that we use a holistic and multi-tactic strategy that includes enhancing the compatibility of pesticides and biological control agents (Hoy 1992; figure 1).

We should preserve sprayed Bt products because they have limited negative effects on the environment, non-target organisms and humans. Sprayed Bt products are especially useful for certain arthropod pests in minor crops, which are increasingly ignored by pesticide companies because they are a small market. Registration of new pesticides for these crops is likely to be more difficult and expensive in the future, which could leave us with extremely limited options for mitigating certain recalcitrant pests. If resistant pests develop in crops containing Bt-toxin genes and they are able to move over to 'minor' crops, then the repercussions of resistance to Bt toxins would be amplified. Sprayed Bt products are limited resources.

An effective paradigm for resistance mitigation has not yet been widely deployed. This is because we have failed to accept that satisfactory resistance mitigation is based on the development of effective, fully integrated multi-tactic IPM programmes. Such programmes ideally will consider the entire agroecosystem and acknowledge the role of monitoring, economic injury levels, biological controls, genetic controls, cultural controls, and biorational controls such as mating disruption, insect growth regulators and mass trapping (figure 1). A key issue in such programmes should always be whether pesticides can be used in a precise and selective manner without disrupting natural enemies. Disruption of natural enemies is not limited to acute toxicity, but can occur if pesticides are applied over a sufficiently large area so that natural enemies are limited in abundance by available food resources. It is time we recognize, as Stern et al. (1959) did, that true resistance mitigation requires a holistic approach to pest management.

This is Florida Agricultural Experiment Station journal series R-06257.

REFERENCES

Barclay, H. J. 1996 Modeling selection for resistance to methods of insect pest control in combination. *Res. Popul. Ecol.* **38**, 75–85.

Brent, K. J. 1986 Detection and monitoring of resistant forms: an overview. In *Pesticide resistance: strategies and tactics for management*, pp. 298–312. Washington, DC: National Academy Press.

Caprio, M. A. & Hoy, M. A. 1994 Metapopulation dynamics affect resistance development in the predatory mite, *Metaseiulus occidentalis* (Acari: Phytoseiidae). *J. Econ. Entomol.* **87**, 525–534.

Caprio, M. A. & Hoy, M. A. 1995 Premating isolation in a simulation model generates frequency-dependent selection and alters establishment rates of resistant natural enemies. *J. Econ. Entomol.* **88**, 205–212.

Caprio, M. A., Hoy, M. A. & Tabashnik, B. E. 1991 Model for implementing a genetically improved strain of a parasitoid. *Am. Entomol.* **37**, 232–239.

Croft, B. A. 1990a *Arthropod biological control agents and pesticides.* New York: Wiley–Interscience.

Croft, B. A. 1990b Developing a philosophy and program of pesticide resistance management. In *Pesticide resistance in arthropods* (ed. R. T. Roush & B. E. Tabashnik), pp. 227–296. New York: Chapman & Hall.

Denholm, I., Devonshire, A. L. & Hollomon, D. W. (eds) 1992 *Resistance 91: achievements and developments in combating pesticide resistance.* London: Elsevier Applied Science.

Devonshire, A. L. & Field, L. M. 1991 Gene amplification and insecticide resistance. *A. Rev. Entomol.* **36**, 1–23.

Flexner, J. L., Westigard, P. H., Hilton, R. & Croft, B. A. 1995 Experimental evaluation of resistance management for two-spotted spider mite (Acari: Tetranychidae) on southern Oregon pear: 1987–1993. *J. Econ. Entomol.* **88**, 1517–1524.

Georghiou, G. P. 1986 The magnitude of the resistance problem. In *Pesticide resistance in arthropods* (ed. R. T. Roush & B. E. Tabashnik), pp. 14–43. New York: Chapman & Hall.

Georghiou, G. P. & Saito, T. (eds) 1983 *Pest resistance to pesticides.* New York: Plenum.

Guillemaud, T., Rooker, S., Pasteur, N. & Raymond, M. 1996 Testing the unique amplification event and the worldwide migration hypothesis of insecticide resistance genes with sequence data. *Heredity* **77**, 535–543.

Hassan, S. A. (and 21 others) 1991 Results of the fifth joint pesticide testing programme carried out by the IOBC/ WPRS-working group 'pesticides and beneficial organisms'. *Entomophaga* **36**, 55–67.

Hokkanen, H. M. T. & Wearing, C. H. 1995 Assessing the risk of pest resistance evolution to *Bacillus thuringiensis* engineered into crop plants: a case study of oilseed rape. *Field Crops Res.* **45**, 171–179.

Hoy, M. A. 1985 Almonds: integrated mite management for California almond orchards. In *Spider mites, their biology, natural enemies, and control*, vol. 1B (ed. W. Helle & M. W. Sabelis), pp. 229–310. Amsterdam: Elsevier.

Hoy, M. A. 1990 Pesticide resistance in arthropod natural enemies: variability and selection responses. In *Pesticide resistance in arthropods* (ed. R. T. Roush & B. E. Tabashnik), pp. 203–236. New York: Chapman & Hall.

Hoy, M. A. 1992 Proactive management of pesticide resistance in agricultural pests. *Phytoparasitica* **20**, 93–97.

Hoy, M. A. 1995 Multitactic resistance management: an approach that is long overdue? *Florida Entomol.* **78**, 443–451.

Hoy, M. A. & Conley, J. 1989 Propargite resistance in Pacific spider mite (Acari: Tetranychidae): stability and mode of inheritance. *J. Econ. Entomol.* **82**, 11–16.

Hull, L. A. & Beers, E. H. 1985 Ecological selectivity: modifying chemical control practices to preserve natural enemies. In *Biological control in agricultural IPM systems* (ed. M. A. Hoy & D. C. Herzog), pp. 103–122. Orlando, FL: Academic Press.

Jaffe, K., Issa, S., Danbiels, E. & Haile, D. 1997 Dynamics of the emergence of genetic resistance to biocides among asexual and sexual organisms. *J. Theor. Biol.* **188**, 289–299.

Lasota, J. A. & Dybas, R. A. 1991 Avermectins, a novel class of compounds: implications for use in arthropod pest control. *A. Rev. Entomol.* **36**, 91–117.

Lewis, W. J., van Lenteren, J. C., Phatak, S. C. & Tumlinson, J. H. III 1997 A total system approach to sustainable pest management. *Proc. Natn. Acad. Sci. USA* **94**, 12 243–12 248.

McKenzie, J. A. 1996 *Ecological and evolutionary aspects of insecticide resistance.* Austin, TX: Academic Press.

Metcalf, R. L. 1994 Insecticides in pest management. In *Introduction to insect pest management* (ed. R. L. Metcalf & W. H. Luckmann), 3rd edn, pp. 245–284. New York: Wiley.

National Academy of Sciences 1986 *Pesticide resistance: strategies and tactics for management.* Washington, DC: National Academy Press.

National Research Council 1989 *Alternative agriculture.* Washington, DC: National Academy Press.

Office of Technology Assessment 1992 *A new technological era for American agriculture.* US Congress, Washington, DC: US Government Printing Office.

Oomen, P. A., Jobsen, J. A., Romeijn, G. & Wiegers, G. L. 1994 Side-effects of 107 pesticides on the whitefly parasitoid *Encarsia formosa*, studied and evaluated according to EPPO guideline no. 142. *Bulletin OEPP* **24**, 89–107.

Pasteur, N. & Raymond, M. 1996 Insecticide resistance genes in mosquitoes: their mutations, migration, and selection in field populations. *J. Hered.* **87**, 444–449.

Pimentel, D. & Lehman, H. (eds) 1993 *The pesticide question. Environment, economics, and ethics.* New York: Chapman & Hall.

Robertson, J. L. & Preisler, H. K. 1992 *Pesticide bioassays with arthropods.* Boca Raton, FL: CRC Press.

Roush, R. T. 1996 Can we slow adaptation by pests to insect transgenic crops? In *Biotechnology and integrated pest management* (ed. G. J. Persley), pp. 242–263. Wallingford, UK: CAB International.

Roush, R. T. & Daly, J. C. 1990 The role of population genetics in resistance research and management. In *Pesticide resistance in arthropods* (ed. R. T. Roush & B. E. Tabashnik), pp. 97–152. New York: Chapman & Hall.

Roush, R. T. & Tabashnik, B. E. (eds) 1990. *Pesticide resistance in arthropods.* New York: Chapman & Hall.

Scott, J. A. 1995 The molecular genetics of resistance: resistance as a response to stress. *Florida Entomol.* **78**, 399–414.

Scott, J. G. 1990 Investigating mechanisms of insecticide resistance: methods, strategies, and pitfalls. In *Pesticide resistance in arthropods* (ed. R. T. Roush & B. E. Tabashnik), pp. 39–57. New York: Chapman & Hall.

Soderlund, D. M. & Bloomquist, J. R. 1990 Molecular mechanisms of insecticide resistance. In *Pesticide resistance in arthropods* (ed. R. T. Roush & B. E. Tabashnik), pp. 58–96. New York: Chapman & Hall.

Stern, V. M. & van den Bosch, R. 1959 The integration of chemical and biological control of the spotted alfalfa aphid. Field experiments on the effects of insecticides. *Hilgardia* **29**, 103–130.

Stern, V. M., Smith, R. F., van den Bosch, R. & Hagen, K. S. 1959 The integration of chemical and biological control of the spotted alfalfa aphid. The integrated control concept. *Hilgardia* **29**, 81–101.

Tabashnik, B. E. 1990 Modeling and evaluation of resistance management tactics. In *Pesticide resistance in arthropods* (ed. R. T. Roush & B. E. Tabashnik), pp. 153–182. New York: Chapman & Hall.

Tabashnik, B. E. & Croft, B. A. 1985 Evolution of pesticide resistance in apple pests and their natural enemies. *Entomophaga* **30**, 37–49.

Tabashnik, B. E., Liu, Y. B., Finson, N., Masson, L. & Heckel, D. G. 1997 One gene in diamondback moth confers resistance to four *Bacillus thuringiensis* toxins. *Proc. Natn. Acad. Sci. USA* **94**, 1640–1644.

Theiling, K. M. & Croft, B. A. 1988 Pesticide side-effects on arthropod natural enemies: a database summary. *Agric. Ecosyst. Environ.* **21**, 191–218.

Wearing, C. H. & Hokkanen, H. M. T. 1994 Pest resistance to *Bacillus thuringiensis:* case studies of ecological crop assessment for *Bt* gene incorporation and strategies of management. *Biocontrol Sci. Technol.* **4**, 573–590.

Index